Mining and its Environmental Impact

ISSUES IN ENVIRONMENTAL SCIENCE AND TECHNOLOGY

EDITORS:

R. E. Hester, University of York, UK
R. M. Harrison, University of Birmingham, UK

EDITORIAL ADVISORY BOARD:

A. K. Barbour, Specialist in Environmental Science and Regulation, UK, **N. A. Burdett,** National Power PLC, UK, **J. Cairns, Jr.,** Virginia Polytechnic Institute and State University, USA, **P. A.** **Chave,** National Rivers Authority, UK, **P. Crutzen,** Max-Planck-Institut für Chemie, Germany, **Sir Hugh Fish,** Consultant, UK, **M. J. Gittins,** Leeds City Council, UK, **J. E. Harries,** Rutherford Appleton Laboratory, UK, **P. K. Hopke,** Clarkson University, USA, **Sir John Houghton,** Royal Commission on Environmental Pollution, UK, **N. J. King,** Department of the Environment, UK, **S. Matsui,** Kyoto University, Japan, **R. Perry,** Imperial College, London, UK, **D. H. Slater,** Her Majesty's Inspectorate of Pollution, UK, **T. G. Spiro,** Princeton University, USA, **D. Taylor,** ICI Group, UK, **Sir Frederick Warner,** SCOPE Office, UK.

How to obtain future titles on publication

A subscription is available for this series. This will bring delivery of each new volume immediately upon publication. For further information, please write to:

The Royal Society of Chemistry
Turpin Distribution Services Limited
Blackhorse Road
Letchworth
Herts SG6 1HN, UK

Telephone: +44 (0) 462 672555
Fax: +44 (0) 462 480947

ISSUES IN ENVIRONMENTAL SCIENCE
AND TECHNOLOGY

EDITORS: R. E. HESTER AND R. M. HARRISON

1
Mining and its Environmental Impact

ROYAL
SOCIETY OF
CHEMISTRY

ISBN 0-85404-200-8
ISSN 1350-7583

A catalogue record for this book is available from the British Library

Published by the Royal Society of Chemistry, Thomas Graham House,
Science Park, Cambridge CB4 4WF

Typeset by Vision Typesetting, Manchester
Printed and bound in Great Britain by The Bath Press, Lower Bristol Road, Bath

Preface

The quality of life in many countries of both the developed and the developing world is strongly affected by the products and processes of mining industries. The first Issue in this new series brings together a set of review articles by leading authorities in the field which, taken together, provide a thorough and detailed survey of many of the key topics of current concern relating to the environmental impact of mining. The level of treatment of most topics is at first introductory but evolves through to the advanced level of current research, backed up by extensive citations of relevant primary literature and other sources. Thus the Issue as a whole will be found equally valuable by students of environmental science and those actively engaged in research related to its theme.

The first article, by A. K. Barbour, provides an overview of non-ferrous metals mining, stressing the importance to modern society of the extractive industries. This touches upon extraction, concentration processes, and environmental impact assessment. The legislative framework is examined and the concept of 'Best Available Technology Not Entailing Excessive Cost' (BATNEEC) is introduced. While the focus of this first article is clearly on the extraction and concentration of non-ferrous metals and the associated environmental impacts, reference is also made to the management of emissions from smelting, to recycling, and to issues associated with the high power requirements of these industries.

Fascinating insights into the many societal and technical concerns associated with gold mining in the Brazilian Amazon are provided by D. Cleary and I. Thornton in the second article. Informal sector mines (garimpos) account for more than 80% of Brazil's gold production, yet it is the world's fourth largest producer. Major environmental impacts result from the extensive use of mercury and the turbidity of rivers resulting from mining operations. This review points out that gold mining is not an important cause of deforestation in the Brazilian Amazon but that the high growth rate of malaria among the local Indian populace is attributable to the mining activity.

The revegetation of metalliferous wastes and land after metal-mining is reviewed by M. S. Johnson, J. A. Cooke, and J. K. W. Stevenson. Many metals, *e.g.* copper and zinc, are essential trace elements at low concentration but become toxic to plants at high concentration, while others, such as lead and mercury, are highly toxic to animals that may graze on the plants. The high acidity and salinity characteristic of mining waste leachates also inhibit plant growth. This article examines the problems and possible solutions related to the improvement of growing conditions and long-term maintenance strategies.

Related issues associated with vegetative remediation at Superfund sites—sites in the USA designated as in urgent need of remedial action after mining—are examined by G. M. Pierzynski, J. L. Schnoor, M. K. Banks, J. C. Tracy, L. A. Licht, and L. E. Erickson. Their article focuses mainly on sites in South Dakota and Kansas. The first is a former gold mining area, where arsenic and cadmium are the principal concerns, and the other is a site where lead and zinc sulfide ores were mined and smelted over a long period of time. The chemical and microbiological aspects of metal-contaminated soils are considered and a model which forms the basis for practical methods of remediation is presented.

D. J. Buchanan and D. Brenkley of the British Coal Corporation review a wide range of issues related to coal mining, including the business environment in which coal must compete, methods of surface and underground production, and current practices to mitigate environmental impact. Their article draws attention to the imperative requirements of controlling the underground workplace environment as well as surveying the rôle of automation and process control, and current and envisaged novel resource exploitation technologies. This is set in an international context but exemplifies UK practice.

The associated problem of methane emissions from coal mining is examined by A. Williams and C. Mitchell. Their article looks at world production in relation to the importance of methane as a greenhouse gas. Knowledge about the emission sources is analysed in terms of three methods of estimation: the Global Average, the Basin or Country Average, and the Mine Specific method. Emissions from both active and abandoned deep mines as well as from surface and post-mining activities are totalled and the technical options for emissions control are evaluated.

Methods for constructing ecosystems and determining their connectivity to the larger ecological landscape are reviewed by J. Cairns, Jr. and R. B. Atkinson. They examine the impact of the 1977 Surface Mining Control and Reclamation Act in the USA which led to careful grading, ditching, and removal of sediment ponds, resulting in rapid dewatering of contour surface mined areas. Their article focuses on recent studies of wetlands in such areas and their design features which relate to industry reclamation needs and regulatory constraints.

The issue of drainage and discharge of polluted water from both active and abandoned mines is examined by R. J. Pentreath of the UK National Rivers Authority. Such water commonly is of high acidity and salinity; iron salts, generating colour in the form of the trihydroxide, constitute a particular problem. Abandoned mines are highlighted: when pumping stops, mines often flood and contamination then enters groundwater and water courses. The legal position is unsatisfactory, as illustrated by the Wheal Jane incident in 1992 when an abandoned tin mine in Cornwall, UK, caused extensive local pollution. The involvement of government agencies in control and remediation measures is described, with other examples drawn from both coal and metals and minerals mining.

Finally, in a review concerned with policy issues and environmental best-practice in metals production, A. Warhurst makes a case for an approach which defines regulatory goals and introduces an informed technology policy. This provides a planning tool to enable regulators to determine optimal corporate environmental trajectories in relation to economic competitiveness and environmental compliance. A number of case studies are outlined with some emphasis on the economic and political considerations.

We would like to thank the RTZ Corporation plc for their kind contribution towards the costs of preparing this publication.

R. E. Hester
R. M. Harrison

Contents

Contents

Contents

Editors

Ronald E. Hester, BSc, DSc(London), PhD(Cornell), FRSC, CChem

Ronald E. Hester is Professor of Chemistry in the University of York. He was for short periods a research fellow in Cambridge and an assistant professor at Cornell before being appointed to a lectureship in chemistry in York in 1965. He has been a full professor in York since 1983. His more than 200 publications are mainly in the area of vibrational spectroscopy, latterly focusing on time-resolved studies of photoreaction intermediates and on biomolecular systems in solution. He is active in environmental chemistry and is a founder member and former chairman of the Environment Group of the Royal Society of Chemistry and editor of 'Industry and the Environment in Perspective' (RSC, 1983) and 'Understanding Our Environment' (RSC, 1986). As a member of the Council of the UK Science and Engineering Research Council and several of its sub-committees, panels, and boards, he is heavily involved in national science policy and administration. He was, from 1991–93, a member of the UK Department of the Environment Advisory Committee on Hazardous Substances and is currently a member of the Scientific Affairs Board of the Royal Society of Chemistry.

Roy M. Harrison, BSc, PhD, DSc (Birmingham), FRSC, CChem, FRMetS, FRSH

Roy M. Harrison is Queen Elizabeth II Birmingham Centenary Professor of Environmental Health in the University of Birmingham. He was previously Lecturer in Environmental Sciences at the University of Lancaster and Reader and Director of the Institute of Aerosol Science at the University of Essex. His more than 200 publications are mainly in the field of environmental chemistry, although his current work includes studies of human health impacts of atmospheric pollutants as well as research into the chemistry of pollution phenomena. He is a former member and past Chairman of the Environment Group of the Royal Society of Chemistry for whom he has edited 'Pollution: Causes, Effects and Control', (RSC, 1983; Second Edition, 1990) and 'Understanding our Environment: An Introduction to Environmental Chemistry and Pollution' (RSC, Second Edition, 1992). He has a close interest in scientific and policy aspects of air pollution, currently being Chairman of the Department of Environment Quality of Urban Air Review Group as well as a member of the DoE Expert Panel on Air Quality Standards and Photochemical Oxidants Review Group and the Department of Health Committee on the Medical Effects of Air Pollutants.

Contributors

R. B. Atkinson, *Center for Environmental and Hazardous Materials Studies, Virginia Polytechnic Institute and State University, 1020 Derring Hall, Blacksburg, Virginia 24061-0415, USA*

K. Banks, *Kansas State University, Great Plains-Rocky Moutain Hazardous Substance Research Center, Department of Chemical Engineering, Durland Hall, Manhattan, Kansas 66506-5102, USA*

A. K. Barbour, *Specialist in Environmental Science and Regulation, 7 Pitch and Pay Park, Sneyd Park, Bristol BS9 1NJ, UK*

D. Brenkley, *British Coal TSRE, Ashby Road, Stanhope Bretby, Burton-on-Trent, Staffordshire DE15 0QD, UK*

D. J. Buchanan, *Director of Research and Scientific Services, British Coal TSRE, Ashby Road, Stanhope Bretby, Burton-on-Trent, Staffordshire DE15 0QD, UK*

J. Cairns, Jr., *Center for Environmental and Hazardous Materials Studies, Virginia Polytechnic Institute and State University, 1020 Derring Hall, Blacksburg, Virginia 24061-0415, USA*

D. Cleary, *Department of Social Anthropology, Cambridge University, Downing Street, Cambridge CB2 3DZ, UK*

J. A. Cooke, *Department of Biology, University of Natal, King George V Avenue, Durban 4001, South Africa*

L. E. Erickson, *Kansas State University, Great Plains-Rocky Mountain Hazardous Substance Research Center, Department of Chemical Engineering, Durland Hall, Manhattan, Kansas 66506-5102, USA*

M. S. Johnson, *Jones Building, Department of Environmental and Evolutionary Biology, University of Liverpool, PO Box 147, Liverpool L69 3BX, UK*

L. A. Licht, *Department of Civil and Environmental Engineering, University of Iowa, 1134 Engineering Building, Iowa City, Iowa 52242, USA*

C. Mitchell, *SPRU, Mantell Building, University of Sussex, Falmer, Brighton BN1 9RF, UK*

Contributors

R. J. Pentreath, *Chief Scientist, National Rivers Authority, Rivers House, Waterside Drive, Aztec West, Almondsbury, Bristol BS12 4UD, UK*

G. Pierzynski, *Department of Agronomy, Throgmorton Hall, Kansas State University, Manhattan, Kansas 66506-5102, USA*

J. L. Schnoor, *Department of Civil and Environmental Engineering, University of Iowa, 1134 Engineering Building, Iowa City, Iowa 52242, USA*

J. K. W. Stevenson, *Group Environmental Scientist, RTZ Limited, 6 St James's Square, London SW1Y 4LD, UK*

I. Thornton, *Global Environmental Research Centre, ICSTM, 56 Queen's Gate, London SW7 5JR, UK*

J. C. Tracy, *Department of Civil Engineering, South Dakota State University, Brookings, South Dakota 57007, USA*

A. Warhurst, *SPRU, Mantell Building, University of Sussex, Falmer, Brighton BN1 9RF, UK*

A. Williams, *Department of Fuel and Energy, University of Leeds, Leeds LS2 9JT, UK*

Mining Non-ferrous Metals

A. K. BARBOUR

1 Introduction

The products of the extractive industries, both metals and minerals, are of pivotal importance to modern life-styles. This situation will continue for the foreseeable future in spite of the inroads made into some non-ferrous applications by plastics, ceramics, and composites. Some of the many applications illustrating this point are indicated in Table 1.

In this introductory review, emphasis is placed primarily on the environmental impacts arising from the mining and concentration of non-ferrous metal ores. Brief reference is made to the efficient management of emissions from non-ferrous smelting processes, recycling, and the environmental issues arising from the significant power requirements of the industries involved.

Unlike organic chemicals and plastics, metals generally cannot be degraded chemically or bacteriologically into simpler constituents, such as carbon dioxide and water, which are relatively neutral environmentally. Metals occur naturally in a wide range of economic concentrations in the ground from approximately 0.05% for uranium, through 0.5–1% for copper, to approximately 60%–70% for iron, and invariably occur in admixture with a wide range of minor and trace metals. Many non-ferrous metals occur naturally as sulfidic compounds. Thus, **metals use is essentially metals relocation** and requires:

(1) *Large energy inputs* to extract the ore and to separate the desired metal from undesired mineral substrates and minor metal impurities, *i.e.* concentration effects.
(2) *Consideration of the toxicity of metals* and associated impurities, *i.e.* their chemical type in extraction, purification, and use (*i.e.* toxicological effects).
(3) *Recycling after use* or, where this is impracticable, permanent disposal in an environmentally acceptable manner, *i.e.* collection and process technology issues.
(4) *Managing the effects of associated impurities*, including associated minor metals and sulfur.

This overall set of processes is summarized in Figure 1.

1

Table 1 Non-ferrous metals are essential to modern society

Housing
- Structural steel protected by galvanizing (zinc)
- Roofing (Architectural and Ancillaries) (lead, zinc, copper)
- Long life windows in aluminium or plastic protected by metallic stabilizers
- Electrical conductors, *etc.* in copper, aluminium

Quality of Life
- Domestic appliance/equipment components die-cast in zinc or aluminium alloys
- Portable tools and appliances powered by nickel/cadmium batteries
- Ornamental items in brass and copper

Transportation
- Automotive batteries (lead/sulfuric acid)
- Car body-shells protected by galvanizing (zinc and zinc–aluminium alloys)
- Electrical equipment (copper and aluminium)
- Stand-by power systems (nickel/cadmium)
- Alloy steels (nickel)

The production, use, and recycling of non-ferrous metals thus requires a complex series of technologies carried out by organizations of widely varying size and sophistication in many areas of the world exhibiting extremes of climate, development, and political outlook.

2 Environmental Background

The desire to protect the environment from the perceived effects of both the extraction and processing industries is strong in the so-called 'developed world' (*e.g.* North America, Europe, Japan, Oceania) and growing rapidly in the 'developing' countries, largely through the efforts of various United Nations agencies. Politicians and regulators express these public wishes through increasingly stringent regulations whose true costs are usually impossible to estimate accurately. Slogans such as the 'Polluter Pays Principle'—whereas the consumer usually eventually pays—are sometimes used to suggest that eventually the costs of building new plants to meet modern environmental standards will become so high that such plants will either not be built or will be constructed in 'developing' countries where standards are thought to be lower.

In general, this view is illusory for new construction and largely so for the upgrading of older plants to modern environmental standards provided an adequate time-scale is allowed; say 5–7 years. It is likely that increasing importance will be attached to environmentally acceptable disposal routes for consumer durable and other end-products. This could result in some market restrictions which would find grudging acceptance from producers and consumers of all environmental standpoints.

The non-ferrous metals industry, in common with its product competitors, has also to manage the impact of quite rapidly rising power costs. Technically, these

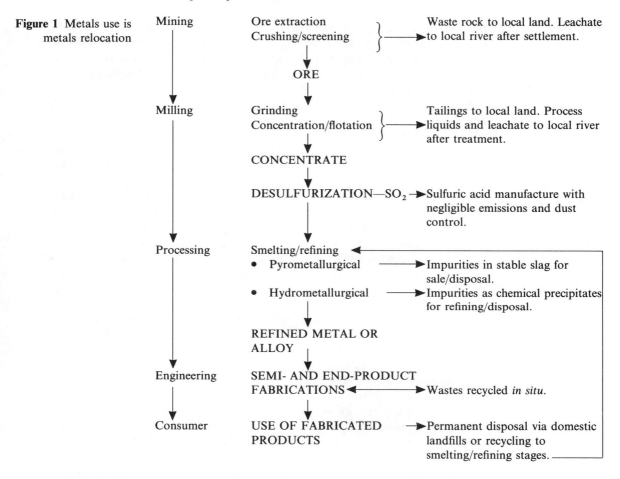

Figure 1 Metals use is metals relocation

increases are attributed mainly to the cost of developing low-sulfur basic sources of energy and the cost of neutralizing acidic emissions at power stations burning coal of relatively high sulfur content, to minimize 'Acid Rain'. The cost of safely decommissioning time-expired nuclear power stations will also become an increasing factor.

Environmental issues are often presented confrontationally—development or environmental devastation; compliance with criteria *versus* costs; industry *versus* the regulators or the 'Greens'—and, indeed, there is never complete congruence between these different viewpoints.

However, the confrontational approach does scant justice to the desires of most people to improve their material standards, not at any cost, but inevitably through industrial activities which provide employment and income as well as products. It also fails to reflect the increasingly general management view that operations must be designed, run, and maintained to the best professional standards, rather than to those which appear to be the most economic in a short-term view.

From a mining and processing standpoint, aspects of implementation of this policy are outlined in the following review. Though mineral extraction, processing, smelting, and refining can never be environmentally neutral, the overall areas of impact are generally quite small. A fully professional approach can achieve a high degree of amelioration provided it is applied consistently and continuously, on a long-term basis, from project initiation to final 'close-out' of the restored and remediated mine and/or refinery.

From the economic standpoint, the cost of meeting inevitably stricter environmental regulations—and the non-regulatory aspects of such disparate issues as accident prevention, including planning for disaster prevention and mitigation, occupational health, product safety, and 'environmental friendliness' in the ultimate end-product—should be judged on a **comparative** basis, relating one product's total cycle costs to those of its market-place competitors. Whilst the future situation *vis-à-vis* competition from plastic and composite materials is much more difficult to estimate with any accuracy, it seems likely that non-ferrous metals will retain many, though not all, applications dependent upon electrical conductivity, ease of repetitive manufacture, and the long-term maintenance of essential physical properties such as strength and relative absence of 'creep' and brittleness. The aesthetic properties of fabricated and well-finished metals will ensure that they are specified for a high proportion of prestige architectural and decorative applications.

Ease and practicability of recycling is already of increasing importance. Unlike metals, most current plastics cannot be recycled without some loss of their original physical properties and so find re-use in less demanding applications. Furthermore, most current plastics are not bio-degradable, *e.g.* in landfills, so that such materials as have to be disposed to landfill can present long-term environmental problems.

Bio-degradable plastics are being developed and, whilst relatively costly at present, plastics may in future be able to add 'environmental friendliness' to their current virtues of relatively easy availability and low finished-item production cost. However, it is virtually impossible to combine bio-degradability with long-term performance in an engineering plastic and, here, metals are likely always to have the advantage, particularly if their relatively easy reprocessing can be exploited in practice to provide higher levels of economic recycling.

General consideration will now be given to the environmental aspects of the separate stages in non-ferrous metals extraction and use.

3 Extraction and Concentration (Mining and Milling)

The production of non-ferrous concentrates can be depicted schematically as in Figure 2.

As noted earlier, natural concentrations of some non-ferrous metals are very low and invariably contain unwanted impurities. Hence, the tonnages of waste products in the form of tailings and overburden can be very large, amounting to many million tonnes per annum from an individual copper or uranium mine. Due to the in-ground concentration effect, tonnages moved and processed are often of the same order for large copper and iron mines. In relation to all foreseen needs, there are ample resources of all metals to be found in the top mile of the earth's crust. The limitations to winning these metals are the availability of cheap power

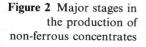

Figure 2 Major stages in the production of non-ferrous concentrates

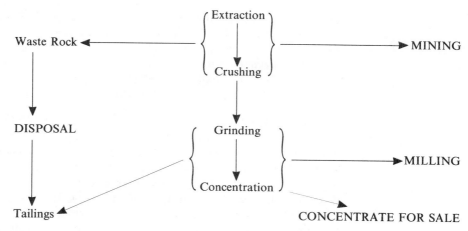

and, to a lesser degree, practicable technology to isolate and extract deeply occurring metals.

Non-ferrous ores are extracted from both open-pit and underground mines, and occasionally from the two in combination. Where a choice is possible from technico/economic considerations, the balance has to be struck between ensuring the health and safety of the miners, usually easier in open-pit than underground mines, and the disposal of waste products, which is usually less intrusive in underground than open-pit mines which have the added problem of 'hiding the hole' at closure. Successful restoration of a worked-out underground mine is usually a simpler task than for an open-pit operation.

Environmental Impact Assessment

Codification and evaluation of all environmental impacts likely to arise from mining and minerals developed is now required in the form of detailed, independent Environmental Impact Assessments by almost all 'developed' and increasing numbers of 'developing' countries before the authorities will grant a licence to proceed. Some of the issues requiring detailed analysis and at least outline ameliorative or mitigation procedures are set out in the following sections.

Location and access. The location of the mine and its ancillaries is usually fixed by the nature of the deposit, though sometimes the mining plan can be modified to take account of particular features, a relatively common one being a feature of great historical or ethnic significance. The locations of the processing plants, intermediate and final product storages, and waste-rock dumps have to be studied with great care, taking account of the historical factors noted above, the restoration/revegetation plan which should be established in outline in the early planning stages, and the minimization of dust-blow from storage piles and conveyors. The areas selected for the deposition of waste rock must not encourage contamination of local streams by run-off nor hinder the restoration plan. The type and location of tailing areas will also justify a major study for all of the above reasons and additional ones, such as dump stability (particularly in seismic areas), rainfall run-off during storms, and dust-blows if high winds occur during arid seasons. The development of suitable and safe access routes to service the mine during both the construction and operational phases is always of vital importance from both the operational and aesthetic standpoints. All of the above factors become of enhanced importance if the operation is located near to significant residential areas or to areas of unusual scientific or ecological value.

Dust-blow. Total elimination of dust arising from blasting, transportation, handling, and storage is impracticable, particularly if the mine is located in an arid area subject to windy conditions. Neither is it practicable to eliminate completely all human activity from the areas generating and emitting dust. Thus many types of amelioration have to be applied, and these include: (1) dampening all areas of dust generation to the maximum practicable extent; (2) paving haul roads at the earliest practicable time, prior to which some chemical treatment or dressing with waste oil are useful temporarily; (3) providing respiratory protection for all exposed workers and ensuring its use; (4) providing mobile equipment operators with a supply of adequately filtered air; (5) ensuring that residential, office, school, and hospital areas are located as far away as possible in areas of minimum dust exposure; (6) covering permanently dumps, conveyors, *etc.* wherever practicable.

Processing operations, particularly crushing and conveying, require specific attention to the design of dust capture and arrestment systems to reduce in-plant dust levels to the relevant standard.

Mine safety. Physical safety standards are always a prime consideration in the design and construction of both open-pit and underground mines and, in most countries, are supervised by a specialist Safety Inspectorate. As in other areas, occupational health standards are correctly being tightened in the light of new information on the effects on health of exposure to contaminants encountered in non-ferrous mining generally; such exposures also include noise and vibration. In general terms, compliance represents a rather small additional cost and, somewhat paradoxically, infractions seem to receive less attention from groups external to the industry, than do environmental issues.

Erosion of Waste-rock Dumps. Unlike the chemical and metallurgical processing industries, mines have to be located where economic mineralization naturally occurs. Since large tonnages of extracted low-value materials have to be transported for upgrading, concentration plant associated with the mine also has to be located nearby. Extraction operations naturally break up the terrain and hence increase greatly the surface area of material exposed to rainfall which, in many parts of the world, falls as intense storms of relatively short duration, giving a high risk of flash flooding.

In these circumstances, 'wash-out' from waste-rock piles is inevitable. Fortunately, by definition, waste-rock contains low concentrations only of the desired elements, which are often relatively toxic, but the clays and silts eroded can cause local streams to become opalescent due to the high burden of suspended solids. Ameliorative measures of general applicability do not exist, though occasionally it is possible to channel run-off streams via the tailings impoundment. Fortunately, a corollary of 'spatey' rainfall is that there are often periods of several months of relatively dry weather when erosion is small and stream discoloration is much less marked. Practical problems arise only where streams subject to serious erosion are used for cattle watering. In these circumstances, provision of alternative supplies of water suitable for the purpose should be provided by the mine operators. This is usually not a particularly onerous requirement since the area of influence of even large, open-pit mining operations is usually quite small and clean supplies can be obtained by the provision of relatively small local impoundments either collecting rainfall or storing the water required for the operations of the milling and processing areas.

Run-off problems can be more serious where sulfidic (pyritic) deposits are being worked or where high-sulfur coal is being extracted. Acid is generated by oxidation reactions:

$$2FeS_2 + 2H_2O + 7O_2 \rightarrow 2FeSO_4 + 2H_2SO_4$$
$$4FeSO_4 + 2H_2SO_4 + O_2 \rightarrow 2Fe_2(SO_4)_3 + 2H_2O$$
$$Fe_2(SO_4)_3 + 6H_2O \rightarrow 2Fe(OH)_3 + 3H_2SO_4$$

Many methods have been proposed for dealing with acidic run-off, including deep injection, neutralization with lime, and dilution. None are of general applicability; such treatments can only be applied where run-off follows well-defined channels, and, in any case, neutralization is both expensive and difficult to operate effectively. Reliance usually has to be placed on the natural absorptive powers of local streams and the ameliorative measures outlined in the preceding paragraph.

In assessing the impact of, and ameliorative measures for, acid generation the following factors would usually require analysis in the Environmental Impact Assessment: (1) location of waste-rock and tailings disposal areas; (2) contribution of each source to the total generated, *e.g.* waste-rock, tailings, mine, processing, *etc.*; (3) practicability of collection by interceptor drains followed by sedimentation, neutralization, *etc.* together with a disposal policy for the solids produced; (4) environmental effects and significance of a no-treatment policy.

Liquid Effluents from Milling.   Milling is the comminution of the extracted ore into particles which can be subjected to a recovery process which separates the valuable materials (concentrate) from the valueless (gangue). The term is now usually used to cover the flotation process (or a chemical treatment process in the cases of alumina production from bauxite; gold; and uranium) which is now an essential part of all non-ferrous mining operations.

After primary and secondary crushing and screening, milling operations start with grinding in a multiplicity of ball and rod mills. After classification, the ground material passes to the flotation units where a variety of reagents may be used, depending on the chemical composition, density, *etc.* of the mineral being concentrated.

Froth flotation, by far the most widely used concentration method, is based on conferring hydrophobicity to the individual particles and hence assisting their attachment to air bubbles. Particles with higher mineral content then rise to the surface of a froth which is skimmed. The remaining barren particles become

Table 2 Flotation Reagents

Class	Use	Compound
(1) Collectors	To selectively coat particles with a water-repellent surface attractive to air bubbles	Water-soluble polar hydrocarbons, such as fatty acids, xanthates
(2) Modifiers		
(a) pH regulators	To change pH to promote flotation; either acidic or basic	$NaOH$, CaO, Na_2CO_3, H_2SO_4, H_2SO_3
(b) Activators and depressants	To selectively modify flotation response of minerals present in combination	Metallic ions, lime, sodium silicate, starch, tannin, phosphates
(3) Frothers	To act as flotation medium	Pine oil, propylene glycol, aliphatic alcohols, cresylic acid
(4) Oils	To modify froth and act as collectors	Kerosene, fuel oils, coal-tar oils

tailings. The flotation reagents used tend to be specific for particular processes. Some general examples are shown in Table 2.

Leaching is the concentration method favoured in some operations, sometimes in conjunction with flotation. The largest-scale example is the separation of alumina from bauxite by the Bayer process in which caustic soda is used to dissolve out the hydrated aluminium oxide; others are the use of sulfuric acid to acid-leach uranium oxide and some copper oxide ores, and the use of sodium cyanide in the extraction of gold.

By passage through a multiplicity of cyclones and thickeners, the end-product of the milling or leaching process is the concentrate of the desired metal or metals, together with a slurry containing the discarded process water, unwanted gangue, and the reagents, frothers, collectors, *etc.* added during the flotation stage. This slurry then passes to the tailings impoundment area, sometimes after chemical treatment, immediately after the flotation section to remove (by oxidation) organic reagents having high oxygen demand, and any cyanides which may be present. Overflow water from the thickeners *etc.*, is recycled back to process wherever possible.

Liquid Effluents from the Tailings Area. In most non-ferrous mining operations, tailings management is a subject of major environmental importance and it requires acceptable solutions to the following issues:

(1) *What is the optimum site for tailings disposal?* In addition to engineering, environmental, and aesthetic acceptability, the site must not infringe areas of historical or ethnic interest and value; nor must it affect the livelihoods of local inhabitants.

(2) *Dam stability and the method of design and construction*, including safe systems for handling exceptional rainfall during heavy storms.

(3) *Tailings stability in seismic areas.*

(4) *Purity of supernatant or run-off water and its disposal route either to recycle or to adjacent streams.*

(5) *The management of any adverse effects which supernatant water may have on adjacent streams and groundwater.*

(6) *The revegetation of tailings areas* to minimize windage losses and to improve aesthetic appearance.

Tailings are inhomogeneous, differing substantially in different non-ferrous mining operations in relation to particle size [slimes (< 200 mesh sieve) to sand(> 200 mesh sieve) ratio], specific gravity, physical characteristics, including abrasiveness, chemical composition, and pH.

The two first parameters influence strongly the flow, settlement, and—in seismic areas—liquefaction characteristics; the chemical aspects naturally have a major influence on the levels of toxicity of the tailings water and treatment methods to minimize its effects on receiving streams or other bodies of water. The details of tailings disposal systems are thus necessarily highly site-specific; the following general outline requires modification and interpretation to suit the details of particular operations in specific locations.

Tailings Disposal—Method and Location. Usually, in mining operations overall, the method and location for tailings disposal has alternative courses of action so that the 'best practicable environmental option' can be selected.

Unlike waste-rock, tailings can be transported as aqueous slurries, either being pumped or moving under the action of gravity through pipes or culverts. Settlement characteristics can be calculated with good accuracy and this is clearly very important for the avoidance of blockages and breakdowns in operation, as is the determination of the degree of abrasiveness on materials of construction likely to be encountered. Slurry transportation often provides a range of options for the economic disposal of tailings which is usually not available for waste rock disposal.

Some such choices which may become practical alternatives are:

(1) *Narrow, deep valleys* versus *disposal in shallow valleys or plain-land*
    Narrow, deep valleys are usually easier to dam and do not disturb agricultural land, although they may obliterate ecologically valuable areas of tropical jungle, *etc.* They are usually visually unobtrusive, partly because they are exposed to the vision of few people. On the other hand, such valley locations are often relatively elevated, thus increasing pumping costs and often increasing hazard in seismic areas if tailings liquefaction ever caused break-out; land at lower elevations is usually more valuable agriculturally.

(2) *Location to minimize adverse environmental impact on adjacent streams, surface waters, and groundwaters*
    For streams and surface waters, the choice of the best practicable environmental option (BPEO) involves weighing and balancing factors such as the degree of treatment (and its cost) required for tailings water disposal into a particular stream *versus* discharging the untreated tailings

into a more distant but environmentally and commercially unimportant stream.

Although now generally out of favour with regulatory authorities, tailings disposal to sea can be a preferred choice in cases where pumping and pipeline costs are not prohibitive. In general, deep outfalls to sea can utilize its enormous absorptive capacity for ions and, usually, the area of serious disturbance to benthic organisms is relatively small. BPEO studies should be made to assist choice between the various options on the basis of both detailed scientific baseline data of all relevant ecological aspects and economics. Although less visually apparent, any adverse effects on groundwater supplies and purity may be very important indeed. Consequently, BPEO studies, based on Environmental Impact Assessments, must include hydrogeologic assessment of seepage flows, *etc.* for the tailings impoundment area, including the dam, as well as the extraction site.

(3) *Location for safety*

The failure of a tailings dam could have disastrous consequences to both human beings and other activities located nearby. Although the design parameters for tailings dams are now well-developed and incorporate safety factors to accommodate predicted frequencies of earthquake and storm, it remains prudent to locate tailings impoundments away from people and human activity as far as possible. This is another clear benefit for sea disposal where it can be done acceptably from the environmental and regulatory standpoints.

Purity of Supernatant Water and Effects on Adjacent Streams.   At most mine sites water is expensive and recycling of tailings effluent is practised wherever possible. At open-pit operations in arid areas, recycled water is frequently used to spray haul roads, broken rock prior to shovelling, *etc.* with the object of suppressing dust to the maximum extent possible.

As noted earlier, the aqueous component of tailings slurry from the mill usually contains very low concentrations of surface-active frothers and collectors and, where acid conditions are present in the flotation circuits, relatively high levels of cations such as iron, manganese, cadmium, mercury, copper, lead, and zinc in specific circumstances. Problems in the tailings area can also be found where pyritic deposits are being worked due to the development of acidity by oxidation in presence of water. When the impoundment is in active use it is usually saturated with water and air access is limited; but when the pond level falls, conditions for rapid development of acidity are present, perhaps posing serious problems with pyritic deposits after operations have formally ceased. Bacterial oxidation with *Thiobacillus ferrooxidans* is thought to be a dominant factor in the development of acidity from sulfur-containing tailings.

If cyanide has been used in the extraction circuit, as in most gold concentration processes, it may be necessary specifically to convert it to relatively innocuous cyanate by oxidation immediately upon leaving the flotation circuits.

It is clearly impractical in this short review to provide worldwide purity criteria for tailings effluent but attention has to be focused on the obvious parameters such as heavy metals (including arsenic) on chloride, sulfate, occasionally fluoride, and, increasingly, nitrate, on suspended solids, and on pH, together with the flow characteristics and uses of the receiving bodies of water. Dependence is usually placed on utilizing the dilution and absorptive powers of the receiving bodies of water. Conventional treatment, *e.g.* liming to precipitate heavy metals, pH adjustment, *etc.* is used where it is necessary to preserve existing uses of the receiving body.

However, in view of the very large volumes of water involved in most tailings operations, particularly where recycling of supernatant is not practised, sludges, *etc.* from treatment processes, usually have to be disposed of separately in small impervious impoundments; this is often not a preferred environmental option compared with dilution into streams as it creates a toxic 'hot-spot' which may be difficult to manage after general operations have ceased. Whatever disposal option is selected, adequate monitoring should be practised to ensure that any significant changes in the quality of the receiving body are quickly detected and assessed.

Revegetation of Tailings Areas and Waste-rock Deposits. During the operating life of the mine, the deposited tailings are normally largely covered by the supernatant mill effluent, leaving only the beaches exposed. This is important for minimizing wind erosion which can become a serious problem where prolonged dry seasons are encountered. At 'close out' or cessation of active operations, it is now becoming usual for regulations to require some permanent system for the management of tailings and waste-rock areas so that they are not a health hazard to either human beings or animals; windage nuisance is minimized, and continued contamination of water courses does not occur. Improvement of aesthetics should also be a significant objective—flat sandy areas can be visually very obtrusive in wooded or mountainous terrain.

Where tailings contain major proportions of slimes, the eventual total 'drying-out' process can be very prolonged and can be accelerated by transpiration from suitable tree plantations. When tailings areas have adequately dried it is often possible to establish vegetation on this barren and hostile substrate using techniques which have developed rapidly over the last 10–15 years. Control of pH by heavy liming is usually a first essential, followed by application of the plant nutrients nitrogen and phosphorus. Grasses, *etc.* indigenous to the area, are often the most promising candidates for successful vegetation. Once a limited natural humus cover has been established, legumes can also be incorporated. Where tailings or waste-rock is highly pyritic, revegetation is much more difficult due to the generation of acid noted earlier, but progress is being made. Of course, all such areas can be top-soiled before re-seeding, but such a procedure is usually inordinately expensive.

Planning for the Avoidance and Mitigation of Disasters. All extraction and processing operations require detailed emergency plans designed to mitigate the effects of major accidents on both the operating personnel and near-neighbours. Both open-pit and underground operations must implement fully all regulatory or professional requirements in relation to physical mine safety.

For neighbourhood protection, close and continuing attention must be paid to the stability of waste piles and tailings areas, particularly dams and retaining walls for tailings disposal areas. All practicable steps must be taken to remove stormwater at an adequate rate and seismic risk must be taken fully into consideration. In the location of tailings areas every effort must be made to choose a location with the minimum possible risk to downstream populations.

Explosives are usually stored in buildings of approved construction and location but it is also vitally important that fuels and chemical reagents are also stored in secure, professionally designed, and bonded (diked) units with written procedures fully implemented for safe loading and discharging from the stores.

Detailed, written emergency plans, including specific responsibilities for identified personnel, must be available and rehearsed thoroughly at regular intervals, normally twice annually.

Site Closure, Remediation, and Restoration. Progressive mine managements support those increasing number of administrations where Impact Assessments require outline closure and remediation plans, usually to be updated as extraction proceeds.

Fundamental to the issue is the optimum location of tailings and waste-rock disposal areas from the standpoint of minimizing environmental impact both during the lifetime of the mine and in the post-closure period. Disposal back into the worked-out pit or underground will generally be impracticable—though some regulatory authorities appear to be thinking in these terms—and so options for the pit itself are restricted to making it secure from trespass with the second option of encouraging or discouraging organized visitors through tourism, depending on the ultimate use of the closed-down operations.

Depending on the weather and hydrology of the area, it may be possible to allow the pit to fill with water, provided it is acceptable for recreational or fishing purposes and does not contaminate local surface or groundwaters. The minerals extraction industries have now developed many leisure complexes, thus providing community value from completed operations.

It is important to store and preserve local topsoil in a biologically active state so that it can be used as a final cover for the waste-rock and tailings areas as they become filled. Such areas will need to be assessed for shaping or 'sculpting' after use. Techniques for improving the aesthetic appearance of such areas by revegetation have made considerable progress in recent years and should always be attempted. Successive managements of UK coal mines and some extraction operations have demonstrated that, with careful planning and management, operational areas can be restored to effective agricultural use. Even if only a low level of vegetation can be persuaded to thrive, this is usually appealing visually and is an important factor in reducing dust-blow, particularly from tailings areas.

Unless a positive decision has been made to develop the worked-out mine as a tourist or educational attraction, the processing buildings, foundations, and contained equipment will have to be dismantled carefully and either sold or disposed of in an environmentally acceptable manner. The inevitable contamination of the plant areas with heavy metals and/or chemical reagents will have to be assessed by specialists and remediated according to their recommendations. A much larger issue, both physically and in terms of ultimate responsibility, concerns the disposal of the 'mining towns', some quite substantial, which have developed, with more or less company participation, near to most significant mining operations. It is outside the scope of this review to do other than note these restoration issues but they are major in scope and not always the subject of clear regulations, particularly as most mines pre-date the requirements of modern Environmental Impact Assessments.

Both public expectation and the professionalism of modern mine managers and operators force the positive conclusion that the local areas of dereliction and the continuing contamination of streams and rivers historically associated with the extraction industries are quite unacceptable today. Whilst the scale of major non-ferrous mining is such that some locally adverse environmental impacts are inevitable during active operations, these can be controlled by active foresight and planning to acceptable levels *for the lifetime of a mining operation*, typically 20–40 years. Techniques actively developed during the last 10–15 years offer considerable promise that long-term dereliction and contamination of river systems can be reduced substantially in the future.

4 Smelting, Refining, and Recycling—Regulatory Developments

Compared with extraction, a larger proportion of these phases of the non-ferrous metals production and use cycle is located in 'developed' countries such as the USA, Europe, Japan, Australia, and the former Soviet Union (FSU). The environmental issues are generally similar to those encountered in the chemical process industries and similar environmental management and control regimes are applied.

In recent years legislative criteria have developed worldwide on the basis of those provided by 'Best Available Technology' (BAT), sometimes, as in the United Kingdom, modified by economic and managerial factors to 'Best Available Technique Not Entailing Excessive Cost' (BATNEEC).

By way of illustration, the UK Environmental Protection Act, 1990, incorporates several new philosophies. Taken together, these will provide a comprehensive system for the control of all process emissions to the external environment to levels which have a rational basis and are as low as can be achieved when modern plants are efficiently operated and maintained. BATNEEC was first embodied in European Community Legislation to control sulfur dioxide emissions and will probably be the basis for future controls promulgated by the European Community. It is also likely to be required as the basis of future projects worldwide supported by international funding agencies such as the World Bank.

The Act will apply the principle of Integrated Pollution Control (IPC) to all processes judged to be of major polluting potential by HMIP (Her Majesty's Inspectorate of Pollution) in the UK. Integrated Pollution Control requires all wastes and emissions to be reduced to the practicable minimum by the use of BATNEEC. Such wastes and emissions as cannot be avoided will be disposed of, as far as possible, using the route causing minimum adverse environmental impact. This will be chosen after considering all options through BPEO studies. It is important to note the use of the word 'Technique' rather than 'Technology' in the UK definition of BATNEEC. 'Technique' includes design and all relevant managerial systems in addition to the technology of the process and its ancillaries.

These principles will be implemented, separately or together, by other regulatory agencies in the UK including the National Rivers Authority (NRA), which is responsible for regulating river quality and estuarial discharges; the Water Services Companies, having responsibility for regulating discharges to sewers; the Local Authority Environmental Health Departments, which deal with the relatively lower polluting-potential operations not regulated by HMIP; and the Local Authority Waste Disposal Units which handle the large tonnages of solid wastes, mainly domestic, which requires permanent disposal in secure landfills or by incineration.

The UK Government has announced that its future plans include the formation of an integrated Environmental Protection Agency from the main Agencies mentioned above to implement the Act and to avoid overlapping responsibilities wherever possible.

The main implications of BATNEEC for the non-ferrous metals smelting and refining operations—and to the major process industries generally—are: (1) the use of Best Available Technology (Technique in the UK) in new plants and the fixed emission criteria which its use implies; (2) the need to submit and obtain authorizations from the relevant Inspectorate. These may be regarded as licences to operate; they will be in the public domain; and will be reviewed regularly, probably at intervals of 3 to 4 years; (3) the upgrading of existing plants to meet current environmental criteria on a more extended timescale, typically 5–7 years; (4) performance monitoring and publication of results.

For the determination of Best Practicable Environmental Option the Inspectorate may require information on matters such as: (1) the process and its relationship with the locality; (2) all emissions leaving the site and the disposal routes that they take; (3) operational data; (4) monitoring information; (5) anticipated effects of significant emissions.

5 Treatment Technologies—Options to Meet Tighter Regulatory Criteria

As noted earlier, the smelting, refining, metal application, and the fabricating/engineering sectors of industry generate significantly different emissions in both type and volume, discharged to a range of media in many different parts of the world.

The metal concentrates produced by the extraction industries for smelting usually contain significant amounts of iron and minor, often toxic, impurities, which consequently have limited markets. For impurities such as cadmium, arsenic, and lead, these markets are reducing further as environmental and health concerns give rise to restrictive legislation and regulatory criteria. Tonnages are considerable, either as by-products from the basic process or arising from the purification of liquid effluents and emissions to atmosphere.

Pyrometallurgical smelters produce siliceous slags in the furnaces which are central to their operation; such slags encapsulate impurities in a form which leaches very slowly and is generally acceptable in well-designed landfills or other disposal areas. On the other hand, hydrometallurgical plants produce the greater part of their solid by-products and wastes in the form of chemical precipitates which are relatively pure and often leachable at a rate dependent upon their chemical and physical properties. Where such materials cannot be sold into the ever-declining markets for them, their ultimate disposal must be to well-designed sealed landfills which require long-term management to ensure environmental security and acceptability.

Lime treatment of liquid effluents produces considerable volumes of material in which metal values are very low. In some processes this material can be recycled for its lime value but, if this is impracticable, disposal to sealed pits is also necessary. Increasingly, 'polishing', using more sophisticated separation techniques, will be necessary to meet tighter criteria.

Platers, anodizers, engineering plants, tanneries, and other operations whose effluents contain non-ferrous metals will also be required to purify them to higher standards prior to discharge into sewer or river. In addition to reducing oxygen demand and adjusting pH, it is likely that processes based on electrolysis, ion-exchange, and reverse osmosis will increasingly be required.

6 Costs

The prolonged recession in developed countries world-wide has caused both industrial managements and some Governmental agencies to appreciate clearly the onerous cost implications of much of the environmental legislation formulated in the prosperous years preceding the recession.

Political slogans such as 'Pollution Prevention Pays', 'Cost–Benefit Analysis', and the 'Polluter Pays Principle' have been shown to be either spurious or of limited applicability. OFWAT, the regulator of the Water Industry in England and Wales, has preceded most other regulators in recognizing that only the consumer can, in the end, pay for the amelioration of pollution, whether it is generated by industry or by the consumption and other activities of consumers. Cost–benefit analysis is applicable to only a few issues and certainly not to the Global questions which are so important in current environmental thinking. 'Pollution Prevention Pays' in a few cases where economic recycling is practicable or where significant process efficiency improvements can be made. In general, however, it has to be recognized that environmental improvement has to be justified on a quality of life and resource basis.

My judgement is that there is no going back on the commitment to use BAT or BATNEEC to produce environmentally acceptable products from modern mines and plants which are designed, operated, and maintained to the best professional standards. Economics may require some delay in the time-scale to achieve BAT but there must be no change in the commitment to achieve the standards it requires.

The Environmental Impact of Gold Mining in the Brazilian Amazon

D. CLEARY AND I. THORNTON

1 Introduction

Despite its recent notoriety, gold mining in the Brazilian Amazon has a long and occasionally distinguished history. Gold was discovered in the south of the Amazon basin, around what today is the city of Cuiabá in the state of Mato Grosso, as long ago as the early eighteenth century. The expeditions of discovery and conquest which followed, spurred by the hopes of finding further gold deposits in the interior, were one of the more important factors behind the formation of present-day Brazil and the driving of its frontiers deep into the South American landmass. The nineteenth century saw a number of attempts by French and British mining companies to establish operations in the Amazon, none of which lasted for more than a few years in the face of the extraordinary logistical difficulties they encountered. The centres of Brazilian gold production by the twentieth century were located outside the Amazon, in the states of Minas Gerais and Bahia, where deep-shaft industrial mining was a more practical proposition. Amazonian gold production played a very minor role in comparison, and was largely ignored by government and mining companies alike until the 1970s.

However, with the construction of the Amazon highway network and the first modern geological surveys of the Amazon, the mineral resources of the region became an important element in development planning, with heavy state and private investment in a number of iron, bauxite, nickel, and cassiterite deposits. Gold mining, on the other hand, followed a very different, less regulated path. The slow rise in the price of gold through the early 1970s stimulated the growth of informal sector mining of alluvial deposits in the Tapajós valley in central Amazonia, drawing on a combination of Guyanese prospecting skills and the development by Brazilian entrepreneurs of new semi-mechanized mining techniques which were well-adapted to river mining. The growth of gold mining in the Tapajós took place independently of the state, and was largely unsupervised and uncontrolled. Then, in 1979, there was an unprecedented rise in the price of gold, which would touch $850 per troy ounce in early 1980. Although it fell back from this peak, oscillating between $300 and $400 in the years to come, this was still an extremely high level, in historical terms, and enough to transform completely the economics of gold mining in Amazonia. Within a very short time, hundreds of thousands of people were attracted from both rural and urban areas into an industry which rapidly became one of the region's most important. This transformation was reflected in the dominant role Amazonian gold production now came to play on the national scene. From 1980 onwards the Brazilian Amazon accounted for at least 75%, and occasionally as much as 90%, of the country's gold production, far outstripping the historical centres of gold production in Brazil. Gold production in the Brazilian Amazon was probably around 100 metric tons (tonnes) annually by 1985, falling off to about 85 tonnes annually in the 1990s. This made Brazil the world's fourth largest gold producer, behind only South Africa, Russia, and the United States, overtaking traditional gold producers like Canada and Australia.[1]

The central fact about Amazonian gold mining is that it has always been dominated by the informal sector. A conservative estimate would be that the informal sector never accounted for less than four-fifths of Amazonian gold production, and often for even more. This will continue to be the case for the foreseeable future. There are several reasons for this, but the most important is that informal sector miners have proved themselves far more capable of responding practically to the special constraints that the Amazon region imposes on mining activity. Amazonia is an extremely expensive area for mining companies to operate within. Distances are huge, transport and energy infrastructure are often either unreliable or non-existent, there is no skilled local labour, and the gold deposits themselves are often spread over wide areas of diffuse mineralization. In these conditions, a deposit needs to be exceptionally high-grade for it to be attractive to a formal sector mining company. Since 1979, an added factor has been that even if a high-grade deposit is located, there are severe problems in preventing the concession being invaded by informal sector miners. Political considerations have often led the Brazilian state to ignore its own mining legislation, or even to intervene against mining companies. In addition to this, while informal sector mining technology in Brazil is simple, it is also appropriate to mining conditions in Amazonia; it has the cardinal virtues of cheapness, portability, and being easy to use, and has proved more effective at locating gold deposits than have formal sector geologists. The result has been to reinforce the dominance of the informal over the formal gold mining sector in Amazonia, to the point where it now seems to be irreversible.

Thus any understanding of the environmental impact of gold mining in the Brazilian Amazon has to begin with the fact that we are dealing with the informal sector, with what in Portuguese are known as **garimpeiros**, informal sector miners, and **garimpos**, informal sector mines. This has a number of consequences. Firstly, there are no reliable figures for any aspect of Amazonian gold mining, only estimates. Secondly, the power of the state to monitor and control what is occurring in garimpos varies from extremely limited to non-existent; this is important when considering actions which might ameliorate environmental impacts, since it implies that such actions need to be taken through co-operation with miners and their organizations, and cannot be imposed from the outside. Thirdly, although the informal sector is often portrayed as chaotic and disorganized—especially by governmental organizations, to which the informal sector, paying no taxes, is by definition illegal—it is in fact highly structured and

[1] D. Cleary, 'Anatomy of the Amazon Gold Rush', Macmillan, London, 1990, Chapter 1, p. 2.

complex. Nobody who has seen a **garimpeiro** operation from close up, or flown over the dozens of airstrips in the remoter corners of Amazonia where garimpos are concentrated, can doubt the high level of organization necessary to mine such large quantities of gold in such a hostile and difficult physical environment. Finally, it should be noted that one of the barriers to a proper understanding of the environmental impacts of Amazonian gold mining has been the notoriety garimpeiros have enjoyed in uninformed media coverage of Amazonia. This has too often filtered through to policymakers. For example, in hearings on the environmental effects of rainforest destruction, the Environment Committee of the House of Commons referred to gold mining as an important cause of deforestation in the Brazilian Amazon,[2] which it is not, and used gross overestimates of the mining population, which appeared to be drawn from journalistic rather than academic or governmental sources.[3] Thus, before the environmental impacts of gold mining can be considered in detail, we need to have an accurate idea of what determines those impacts—the scale and location of the mining, the technologies involved, and the trading process.

2 The Amazon Gold Rush Since 1979

Amazonian gold mining, like many informal sector activities, could have been designed to make the statistician's task impossible. There are no reliable production figures. Production estimates are deduced from the volume of sales in the domestic Brazilian gold market and from reports by the federal government's mining agency, the Departamento Nacional da Produçao Mineral (DNPM), which maintains field stations in the most important goldfields. The reliability of these estimates varies considerably, and is also affected by the fact that much Brazilian gold is smuggled into neighbouring countries, making the domestic Brazilian gold market only an approximate indicator of the true level of mining activity. Nevertheless, these estimates are important because knowing the level of gold production is the first step in calculating the amount of likely mercury pollution, the most serious environmental consequence of informal sector mining in the Amazon. If gold production was 100 tonnes in a year, for example, this means that approximately 100 tonnes of mercury were probably released as mercury vapour in gold trading areas, and that somewhere around 100 tonnes of mercury was washed into the Amazonian river system as spillages from mining equipment during that year. Brazilian researchers have suggested this may be equivalent to 1% of total global emissions of mercury into the atmosphere, and 6% of anthropogenic emissions, with spillages into the river system being approximately equal to the annual mercury input into the North Sea.[4]

Most estimates suggest Amazonian garimpos have produced between 80 and 100 tonnes of gold annually since the early 1980s. Calculating the population of garimpos is even more difficult, since it fluctuates according to time of year—garimpo areas reduce operations during the rainy season—and economic

[2] Environment Committee, Session 1990–91, 'Climatological and Environmental Effects of Rainforest Destruction', Third Report, HMSO, London, 1991, p. 101.

[3] Environment Committee, Session 1989–90, 'Report of a Visit to Brazil', HMSO, London, 1990, p. 10.

[4] W. Pfeiffer and L. Lacerda, *Environ. Technol. Lett.*, 1988, **9**, 328.

Map 1 Principal goldfields in the Brazilian Amazon

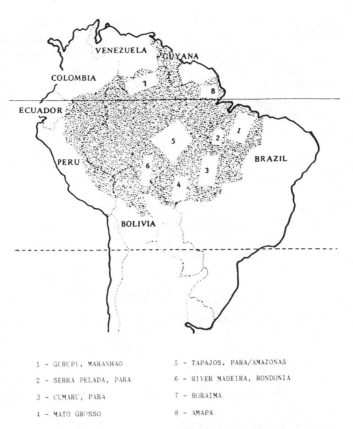

1 - GURUPI, MARANHAO	5 - TAPAJOS, PARA/AMAZONAS
2 - SERRA PELADA, PARA	6 - RIVER MADEIRA, RONDONIA
3 - CUMARU, PARA	7 - RORAIMA
4 - MATO GROSSO	8 - AMAPA

conditions in the areas of origin of the workforce. A reasonable estimate would be that approximately 300 000 garimpeiros work in the Brazilian Amazon. They are concentrated in nine major goldfields, with the most important, in terms of numbers and production, being along the Tapajós and Madeira rivers (see Map 1).

The technologies used in the mining process can be either manual or mechanized; a feature of informal sector gold mining during the 1980s has been the displacement of manual methods by mechanized techniques. The oldest manual technologies still found in Brazilian garimpos are **lontonas** and **dallas**, which are portable wooden sluices through which a mixture of alluvium and water is poured. These are the same techniques as were used in the nineteenth century gold rushes, as photographs of mine workings in Australia and California from that period show,[5] and the words lontona and dalla are clearly Portuguese versions of the 'long tom' and the 'dollar' sluices which were used by nineteenth century gold miners. These appear to have been introduced to Brazil in the 1940s by miners moving down from the then British Guiana. The Brazilians in turn produced their own form of sluice called a **cobra fumando**, still common in Amazonia, which consists of a tin filter set in a wooden box with a

[5] D. Lavender, 'The American West', Penguin Books, Harmondsworth, 1969 and D. Stone, 'Gold Diggers and Gold Digging: A Photographic History of Gold in Australia, 1854–1920', Lansdowne Press, Melbourne, 1974.

small sluice attached, through which the alluvium is scrubbed. The sluices are lined with sacking across which riffles are laid, and mercury is poured behind the riffles to amalgamate with gold particles which will then lodge in the sacking. Mercury is also often introduced into a box at the top of the sluice where the filtered material enters.

There are four main forms of motorized mining technology in garimpos; in ascending order of sophistication they are the **moinho**, the **chupadeira**, the **balsa**, and the **draga**. The moinho is a small mechanical crusher used for working both gold ore and alluvium; it consists of two steel flails in a metal casing, driven by a fanbelt operated by a small engine. A sluice is attached and mercury used in exactly the same way as in manual technologies. More sophisticated ore-crushing technologies, such as stamp mills and ball crushers, are unknown in Brazilian garimpos. The chupadeira consists of a high-pressure water hose, which is used to dislodge alluvium or friable ore, which is then passed over a large sluice by a motor suction pump. A balsa is essentially a chupadeira mounted on a raft, with the water hose being used on a river or stream bed by a miner in full diving gear. A draga is as sophisticated as garimpo technology gets: it consists of a jointed metal tube mounted on a large raft with custom-built metal hulls. The tube ends with a drill bit and a hydraulic pump. The tube is raised and lowered by steel cables powered by a truck engine mounted on the hulls, and the alluvium from the river bed is passed over large sluices on the raft. Dragas differ from other garimpo technologies in that no mercury is used in the sluices; the gold is separated from sediments by collecting the material from the sluice lining and washing it in an oil drum, where the mercury is introduced, with separation then taking place by manual panning. This means dragas spill much less mercury than balsas or chupadeiras, where mercury is used in the sluices. Dragas are used to mine gold on large rivers such as the Madeira, where they are the dominant mining technology. Chupadeiras and balsas are used on creeks and smaller rivers, and are the dominant technology in the Tapajós valley. Although there are no records of amounts of gold mined by these different technologies, there is little doubt that the most widespread are the chupadeira and the balsa. These happen to be the two technologies which use the most mercury in the sluice, and where the nature of the work is such that the highest rate of direct spillage of mercury into aquatic ecosystems can be expected during mining operations, compared to other mining technologies found in garimpos.

Mercury is an integral part of mining technology in the garimpo. Where chupadeiras are being used, mercury is often sprinkled over the area to be excavated before the hoses are turned on it, if gold levels are believed to be reasonably high. The common feature of all mining technologies in the garimpo, save for the draga, is the use of mercury on a sluice. Particles of gold amalgamate with mercury, and this makes it more likely that the gold will be retained in the lining of the sluice. If the machinery is being operated perfectly, a certain volume of gold will amalgamate with an equal volume of mercury, and surplus mercury will be collected and recycled when the apparatus is cleaned at the end of the work period. Of course, this does not always occur. Mercury is frequently washed away because water velocity over the sluice is too high, or it may be spilt as it is being put into the machinery, or when the machinery is being cleaned. Inevitably, then, it becomes very difficult to quantify the amount of mercury being lost in spillages; it is probable that it is at least equivalent to the volume of gold production.

After mercury–gold amalgam is collected from the mining machinery, it then undergoes a series of burnings which are linked to the gold's progress along a trading network which will take it from the mining site to a refined ingot of pure gold anywhere from São Paulo to Geneva. This sequence of burnings needs to be kept in mind when assessing the likely impact of the mercury vapour released as a by-product, since the type of burning which takes place in the location where the gold is extracted and those which occur as it moves along the trading chain is significantly different. Usually, gold mined in Amazonian garimpos is subjected to three separate burnings before it is finally refined into pure gold. The first takes place at the actual site of extraction, carried out by the miners themselves, usually in the open air, when the mercury–gold amalgam is placed on a metal pan and burnt using a butane torch. The awkwardness of moving the gas cylinders means that repeated burnings take place in the same spot. Also, as the cylinders are used to fuel gas cooking rings, it often takes place in an area of food preparation. It is at this stage that most mercury vapour is released. As the gold cools, it acquires its familiar colour, but will still contain a certain proportion of mercury, along with other impurities such as iron. This impure gold is then taken to a trading centre; most often this will be the nearest garimpo to the site of extraction. Here, the gold receives a second burning, since the buyer pays by weight and is obviously anxious to ensure that he does not pay for impurities the gold contains. This second burning may take place in the open air, on the street in front of the trading post, or in crude home-made fume cupboards inside the post. No masks or other protection are used, nor are filters usually fitted to the fume-cupboards; vapour is typically piped straight out into the street, or into the narrow alleys between houses. Although the volume of mercury vapour released in the second burning is lower than in the first, as traders may buy gold many times a day, the greater volume of gold being burnt means that traders, rather than the miners themselves, are the social group most likely to be at risk from contamination through the inhalation of mercury vapour.

Gold traders in garimpos in turn sell their gold on to large gold-buying concerns. They in turn will burn the gold before paying for it, to ensure they also are not paying out for impurities. Thus, in cities which occupy a strategic place in the gold economy, such as Itaituba, Porto Velho, Santarém, and Cuiabá, a concentration of major gold traders will be burning quite large amounts of gold: in the case of Itaituba, for example, probably between 12 and 20 tonnes of gold annually. However, the gold they burn will typically already have been burnt twice, and the level of mercury it contains will be relatively low, which suggests the danger to the urban population in gold trading towns may not be as great as the amount of gold being traded might suggest. Finally, once the gold arrives in the hands of a major gold trader, it will be refined in a sophisticated smelter into pure gold ingots. There are two smelters within the Amazon, in the town of Itaituba, and other smelters outside the region, in Rio and São Paulo. Gold smelting, even within Amazonia, requires sophisticated technology in which mercury is condensed and recycled, minimizing the pollution risk.

3   The Environmental Impact of Garimpo Mining

Observers concur that the most serious potential consequence of the modern Amazonian gold rush is the uncontrolled use of mercury. However, the description of mining techniques above should make it clear that mercury use is not the only environmental problem associated with mining. Given that the relationship between, for example, informal sector mining and deforestation has been controversial, it is proposed in this section to deal with non-mercury related environmental impacts. The mercury issue will then be dealt with at length in the concluding section of this chapter.

Mining and Aquatic Ecosystems

One of the most visually striking consequences of gold mining in the Amazon is the change that it provokes in aquatic ecosystems. All mining techniques used in Amazonian garimpos disturb river and stream sediments, increase siltation rates, and may lead to radical changes in aquatic life. Mining operations can pollute streams and rivers through the uncontrolled run-off of tailings, spillages of petrol or diesel oil, and detergents. Detergents are used in areas of fine-grained gold to prevent the formation of micro-bubbles which might attach themselves to fine-grained gold particles and allow them to be flushed away rather than retained in machinery. Most far-reaching in its effects, however, is the widespread disturbance of river and stream sediments which are the inevitable consequence of the widespread use of balsas, dragas, and chupadeiras. They increase turbidity to the point where miners working underwater from balsas are literally unable to see their hands in front of their faces, the level of suspended particles increases to the point where streams and rivers have their colour altered to a muddy brown, and there are drastic declines in fish populations. The appearance of large shoals of dead fish in Mato Grosso and Pará states, near garimpos, has been linked, by local Brazilian researchers, to mining; although no detailed studies have yet proven a link between garimpos and fish mortality, it is difficult to see what else might be responsible. It is certainly the case that the river Crepurí, a tributary of the Tapajós, has changed colour from a clear water river in the mid-1970s to its current muddy brown as a direct result of the use of balsas. Perhaps the best-documented case is that of the Rio Fresco, in the Gorotire reserve of the Kayapó Indians in the south of Pará state. Before the early 1980s, this was a crystalline river which the Kayapó used for fishing and drinking water. After the discovery of gold at the borders of the reserve in 1982, and the growth of gold mining upriver, the river turned opaque, fish yields declined, and the Indian agency, FUNAI, found it necessary to pipe water into Gorotire village as the river water was no longer fit to drink. The fact that fish is a staple of the diet of riverine populations in the interior of the Amazon means that alterations in the fish population due to mining—perhaps a change in fish migration patterns, for example—may have unforeseen consequences for riverine communities even beyond the immediate area affected by mining.

Mining and Malaria

One of the most worrying changes in disease patterns in recent years in the Amazon basin has been a resurgence of malaria, after a period when it seemed to be in decline as a result of malaria control activities. Between 1970 and 1989, reported malaria cases in Brazil increased from 52 469 to 577 520; virtually all autochthonous malaria in Brazil is now concentrated in the Amazon region.[6] Garimpos have been identified as the form of human settlement in the Amazon associated with the highest levels of malaria transmission.[7] The excavations made by miners, which fill with rainwater, form ideal mosquito breeding grounds, and the mobility of infected miners has greatly aided the spread of *P. falciparum*, the most dangerous of the three malaria plasmodiae found in the Brazilian Amazon. This can have a particularly serious effect on Brazilian Indian populations, who typically possess antibodies for *P. vivax* and *P. brasilianum*, but may suffer greatly from *P. falciparum* in the early stages of contact. Most data exists for the Yanomami Indians of northern Amazonia, who have been affected by garimpeiro incursions on a large scale since the mid-1980s. In 1990, 60% of deaths among the Yanomami were attributed to malaria, and some reports suggest that 15% of the Brazilian Yanomami died of malaria in only three years in the late 1980s.[8] In established garimpo areas such as the Tapajós and Madeira rivers, the juxtaposition of new arrivals and symptomless old-timers who act as reservoirs of gametocytes results in repeated infection, with many miners losing count of the number of malarial attacks they have suffered, so regular does reinfection become.

Mining and Deforestation

Contrary to the beliefs of many observers, garimpos make an insignificant contribution to levels of deforestation in the Amazon region. This is not to say that no deforestation takes place in mining areas; miners will regularly clear a hillside of vegetation if they are following a vein, and some deforestation occurs as a result of the cutting of forest paths between work fronts, or to facilitate mining operations on the forested banks of a creek or river. The clearing of airstrips through which many garimpos are supplied also often involves the clearing of forest. Nevertheless, when garimpo mining is compared to other land uses in the Amazon which also involve clearing forest, such as ranching, smallholder agriculture, or logging, the amounts of forest cleared by garimpeiros are insignificant. This is particularly noticeable when flying over the Tapajós goldfield. Despite the fact that garimpo mining has been occurring on some scale in the Tapajós since the late 1950s, the forest areas around the older garimpos are still largely intact. It is perfectly possible to walk into primary forest, which still stands undisturbed, a few minutes from a garimpo which is over thirty years old,

[6] D. Sawyer, 'Malaria and the Environment', Working Paper No. 13, ISPN/Inter-Regional Meeting on Malaria, Pan American Health Organization, Brasilia, 1992, p. 2.

[7] D. Sawyer, 'Malaria and the Environment', Working Paper No. 13, ISPN/Inter-Regional Meeting on Malaria, Pan American Health Organization, Brasilia, 1992, p. 15.

[8] D. Sawyer, 'Malaria and the Environment', Working Paper No. 13, ISPN/Inter-Regional Meeting on Malaria, Pan American Health Organization, Brasilia, 1992, p. 16.

such as Cuiú-Cuiú. Such deforestation as exists in the Tapajós has occurred near the regional trading centre of Itaituba, some hundred kilometers from the nearest garimpos, and is associated with ranching and agriculture, not mining. The same is true of the Madeira goldfield. Although the state of Rondônia, where the Madeira goldfield is located, suffered the highest levels of deforestation anywhere in the Amazon during the 1980s, all researchers agree that the forest was cleared by ranchers, smallholders, and loggers, and that the garimpeiro contribution to Rondônian deforestation was negligible.

4   The Impact of Mercury Use in the Brazilian Amazon

Over recent years, extensive environmental contamination has occurred in certain areas of the Brazilian Amazon as a result of gold recovery processes. Mercury–gold amalgam, when heated, releases elemental mercury vapour. This may give rise to high atmospheric contamination at the worksite and at trading posts, with absorption of mercury by inhalation a possibility not only for miners and traders, but other persons in the vicinity of the burning site. The deposition of mercury is likely to have given rise to raised mercury levels in soils, with subsequent uptake by plants, and raised mercury levels in sediments and waterways. In the aquatic environment methylation of mercury into a more toxic organometallic form occurs in part through microbial action, with methyl mercury entering the food chain by tight binding to proteins, resulting in biomagnification, with the highest levels to be expected in predatory fish species. The contamination of waterways with elemental mercury leads to increased uptake and absorption of methyl mercury in a fish-eating community. Account also should be taken of the global cycle for mercury in which elemental mercury vapour released from amalgam processing is converted to soluble forms and deposited by rain into soil and water.[9]

The pattern of adverse health effects that mercury may provoke is determined by its speciation. Elemental mercury vapour is readily and rapidly absorbed by inhalation and deposited in target tissues, in particular in the central nervous systems and the kidney. Methyl mercury is almost completely absorbed by ingestion, and is distributed through the blood to all tissues, with maximal deposition in the brain. Elemental mercury is mainly excreted via the kidney into the urine. Methyl mercury is excreted into the bile and via faeces. It is in part converted in the body to mercuric mercury. Following exposure to mercury, the initial symptoms of poisoning are erethrism and tremor, which may be difficult to distinguish from similar symptoms caused by tropical disease. Severe contamination causes narrowing of the visual field, seizures, coma, and death.[10]

Any discussion of the implications of mercury use in Amazonia should begin with an acknowledgement of the limits of our knowledge of the real situation. Despite the widespread attention that the topic has received in the media and elsewhere, the central problem in assessing the impact of current levels of mercury

[9] I. Thornton, N. Brown, D. Cleary, and S. Worthington, 'Mercury Contamination in the Brazilian Amazon: A Report for the Commission of the European Communities', Directorate-General 1-K-2 Environment, Contract Reference B946/90, Brussels, 1991, p. 39.

[10] WHO, 'Environmental Health Criteria 101—Methylmercury', WHO, Geneva, 1990.

use on the Amazonian environment, and local populations, is the lack of large-scale environmental and epidemiological survey results. It may seem surprising that such information is still hard to come by. But the reasons are straightforward enough: the expense and logistical problems of this kind of research, given the lack of appropriate laboratory facilities within the Amazon region itself, the severe financial crisis affecting research institutes in Brazil, and the need to translate complex findings out of Portuguese into English in order to publish in international journals. Much work remains inaccessible to non-Portuguese speakers, buried in government and research institute reports. There is the further problem that outside the large cities of Amazonia very little epidemiological information of any kind exists, let alone the detailed data on foetal abnormalities and peri-natal mortality at the sub-municipal level that would be necessary to make a satisfactory assessment of the true level of health risk that mercury poses in the region.

Nevertheless, a number of mainly Brazilian researchers have managed to overcome these problems and publish important findings in the international scientific literature.[11] Virtually all published research examines mercury concentrations in the environment (air, plants, soil, river sediments) in and around mining and gold trading areas. There is some information on fish,[12] but only one published source in the international literature to date on mercury concentration among the human population, an examination of hair mercury levels among 34 miners.[13] We have obtained data from 106 people from the Tapajós valley which will be published shortly. Nevertheless, the scarcity of reliable human data makes it very difficult to assess the real impact of mercury use on public health in Amazonia at present, although there are suggestive indications from the levels of mercury found in fish. Commonly eaten fish species from the Madeira river, where fish is the dietary staple of a large riverine population, were found to contain elevated levels of 1.01 ± 0.64, 0.13 ± 0.08, and $0.12 \pm 0.06\,\mu\mathrm{g\,g}^{-1}$ for carnivorous, omnivorous, and detritus-feeding species, respectively.[14] This means that a daily intake of between 10 and 20 g of carnivorous fish by a child weighing 20 kg can easily result in mercury poisoning, as the WHO recommended tolerable daily intake limit is $0.43\,\mu\mathrm{g\,kg}^{-1}$ body weight.[15] Such daily dosages would be routinely exceeded in most riverine villages.

[11] J. Andrade, M. Bueno, P. Soares, and A. Choudhuri, *An. Acad. Bras. Cienc.*, 1988, **60**, 293–303; L. Lacerda, W. Peiffer, A. Ott, and E. Silveira, *Biotropica*, 1989, **21**, 91–93; L. Lacerda, F. DePaula, A. Ovalle, W. Pfeiffer, and O. Malm, *Sci. Total Environ.*, 1990, **97/98**, 525–530; O. Malm, W. Pfeiffer, C. Souza, and R. Reuther, *Ambio*, 1990, **19**, 11–15; L. Martinelli, J. Ferreira, R. Victoria, and B. Forsberg, *Ambio*, 1988, **17**, 252–254; J. Nriagu, W. Pfeiffer, O. Malm, C. Souza, and G. Mierle, *Nature (London)*, 1992, **356**, 389; W. Pfeiffer and L. Lacerda, *Environ. Technol. Lett.*, 1988, **9**, 325–330; W. Pfeiffer, L. Lacerda, O. Malm, C. Souza, E. Silveira, and W. Bastos, *Sci. Total Environ.*, 1989, **87/88**, 233–240.

[12] O. Malm, W. Pfeiffer, C. Souza, and R. Reuther, *Ambio*, 1990, **19**, 11–15; L. Martinelli, J. Ferreira, R. Victoria, and B. Forsberg, *Ambio*, 1988, **17**, 252–254; J. Nriagu, W. Pfeiffer, O. Malm, C. Souza, and G. Mierle, *Nature (London)*, 1992, **356**, 389; W. Pfeiffer and L. Lacerda, *Environ. Technol. Lett.*, 1988, **9**, 325–330.

[13] O. Malm, W. Pfeiffer, C. Souza, and R. Reuther, *Ambio*, 1990, **19**, 12.

[14] J. Nriagu, W. Pfeiffer, O. Malm, C. Souza, and G. Mierle, *Nature (London)*, 1992, **356**, 389.

[15] 'WHO Environmental Health Criteria 101: Methylmercury', World Health Organization, Geneva, 1990.

In the Tapajós, an Anglo/Brazilian research team recorded levels as high as $2.575\,mg\,kg^{-1}$ wet weight in fish samples, and 21 of a total of 51 fish samples were found to exceed the European Community Environmental Quality Standard of $0.300\,mg\,kg^{-1}$ for a 'basket of fish'.[16] The same research team found levels of mercury in blood and urine which suggested some gold traders in garimpos, and some fish-eaters in the riverine village of Jacareacanga, some distance from the closest mining area, had a high probability of experiencing adverse health effects. European Community occupational health guidelines suggest that blood contamination is an indicator of recent exposure to elemental mercury, while urine concentration is an indicator of renal accumulation, for example, in the case of inhalation of mercury vapour. In the case of urine, removal from exposure is recommended at levels greater than $50\,\mu g\,g^{-1}$, and in the case of blood removal is recommended at levels greater than $10\,\mu g\,(100\,ml)^{-1}$. It is clear from our recent studies that traders and fish-eaters may be the groups most seriously affected by mercury exposure. The levels of blood mercury in the fishing village of Jacareacanga are a particular cause for concern, given its distance from areas of mercury use, and the importance of fish in the local diet. It seems likely that the larger population and higher fish mercury concentrations found in the river Madeira mean that the health implications of mercury pollution may be worse there than in the Tapajós valley.

On the question of mercury contamination in the environment, the literature is also suggestive, rather than comprehensive. It extends a long way beyond the areas of actual mining, with data from the Madeira river suggesting most of the mercury is transported some 500 km from the garimpo areas to the main channel of the Amazon.[17] Similar published research has not yet been carried out along the Tapajós river. Nevertheless, elevated mercury contamination in fish taken along the main channel of the Amazon has never been recorded in the occasional sampling carried out by state bodies and researchers. Along the Madeira river system mercury concentrations found were variable, with the highest levels coming from tributaries. Concentrations have also been recorded in some river waters, bottom sediments (some in the Madeira as high as $19.8\,\mu g\,kg^{-1}$),[18] and river plants.[19]

As this brief summary of the research literature shows, a severe problem in assessing the impact of mercury use in the Amazon region as a whole is the bias of the published research towards the Tapajós and Madeira goldfields. While these are the two areas where the most mercury has been used for the longest period, and are therefore where one might expect the impacts to be most severe, between them they contain less than half of the mining population, and less than half of the population likely to be affected by mercury use. Even within the Tapajós and Madeira, there is no published information about the impact of mercury on such

[16] I. Thornton, N. Brown, D. Cleary, and S. Worthington, 'Mercury Contamination in the Brazilian Amazon: A Report for the Commission of the European Communities', Directorate-General 1-K-2 Environment, Contract Reference B946/90, Brussels, 1991, p. 31.

[17] J. Nriagu, W. Pfeiffer, O. Malm, C. Souza, and G. Mierle, *Nature* (*London*), 1992, **356**, 389.

[18] W. Pfeiffer, L. Lacerda, O. Malm, C. Souza, E. Silveira, and W. Bastos, *Sci. Total Environ.*, 1989, **87/88**, 233.

[19] L. Martinelli, J. Ferreira, R. Victoria, and B. Forsberg, *Ambio*, 1988, **17**, 253.

important groups as Indians—there are indigenous reserves near mining zones in both areas—nor do we have any reliable information on, for example, foetal abnormalities. Our knowledge of the situation within the two best-studied areas of garimpo mining still leaves much to be desired: in other important gold mining areas, such as northern Mato Grosso, the Yanomami area of Roraima, and the Cumarú goldfield in southern Pará, no data at all have been published. Where ignorance reigns, speculation flourishes. The Yanomami, for example, are regularly and wrongly reported in newspapers as suffering from large-scale mercury poisoning when it is quite clear that the malaria introduced by garimpeiros is an infinitely more important threat to them, given that fish is not an important part of the Yanomami diet. It is dangerous to extrapolate what little we do know of the Tapajós and the Madeira to other mining regions, where the technologies, local environment, and river systems may all differ from the Tapajós and Madeira.

Nevertheless, enough is known about mining technologies, the scale of mercury use, the location of goldfields in relation to the regional population, and the nature of the trading process, to make some general concluding remarks. Firstly, there are a number of possible contamination pathways into the human population, which fall under the two general headings of mercury spillages from the mining process, and the release of mercury vapour during the trading process. The most potentially serious in public health terms is mercury spillages, since there is the clear danger that mercury can enter the aquatic food chain, and affect a large riverine population in the interior of the Amazon, both rural and urban, whose diet revolves around fish. It may also contaminate other aquatic life important to the regional diet in certain areas, such as crabs, turtles, and turtle eggs. The data available suggests mercury contamination is high enough for this to be occurring at least along some stretches of the Tapajós and Madeira rivers. Other areas of gold mining where the river systems are comparatively smaller, where balsas are not so important a method of gold extraction, and fish not so important an element of the regional diet, will be less affected: goldfields where these attenuating factors hold good are the Gurupí region, Roraima (the Yanomami area), and, to a lesser extent, Amapá. Mato Grosso and Cumarú present a similar risk profile to the Tapajós and Madeira goldfields, but the fact that gold production began considerably later in the former compared to the latter means that the situation may not be as serious. The risk of contamination through inhalation of mercury vapour will be greater in the nodes of the trading system where the burning takes place: the more prosperous garimpos, and gold trading towns and cities. However, as explained, the nature of the trading process is such that gold being burnt in the major trading centres has a lower level of mercury, and those most at risk are probably the small and medium-sized traders operating in garimpos. Besides inhalation of vapour, there are other possible contamination pathways in trading centres. Mercury spillages may occur wherever it is stored, such as a trading post, and the young children of a gold trader could, for example, ingest mercury particles while crawling in or around the trading post.

Knowing what we do of the social composition of mining areas and the goldfields in general, and the pattern of mercury use and trading, we can also define the groups of the local population most likely to be at risk. Miners, traders, and those who consume a great deal of carnivorous fish are clearly vulnerable. However, there is a great deal of variation in Amazonia in fish-eating habits, with carnivorous species being preferred in one location and non-carnivorous in another, so even with enthusiastic eaters of contaminated fish we can expect significant variation in the physical manifestation of symptoms, with some who might be expected to be at risk either not showing any symptoms of mercury contamination, or only sub-clinical symptoms. Among the fish-eating population, pregnant women are a particularly important risk group, since the foetus is particularly vulnerable to methyl mercury. The lack of any data for infant mortality or foetal abnormalities in riverine communities makes it impossible to assess the level of this threat at present. The situation is further complicated in many parts of Amazonia by cultural beliefs which ascribe spontaneous abortions and deformities to supernatural causes, which makes affected women reluctant to discuss them with outsiders. Lengthy preliminary work would therefore be necessary in such communities if reliable information on this critically important topic is to be gathered.

In conclusion, it is worth pointing out that no likely alternative to mercury use exists in the informal gold mining sector in Brazil. Leaching technologies are too expensive and too operationally complex for them to be an option to any but a tiny minority of Amazonian garimpeiros. While some remedial actions are possible in the case of mercury vapour, such as the fitting of charcoal filters to fume cupboards and the use of purpose-built retorts, the weakness and corruptability of public authority in much of Amazonia make even this problematic. Given the number of livelihoods which depend on gold mining and its support industries in the interior of the Amazon, no politicians are likely to press for any regulation of the gold trade, unless a serious public health risk can be demonstrated in much more detail than has yet been possible. Any ban on the importation of mercury, or the mercury trade, would be counter-productive. The means for policing such a ban do not exist, and the certain consequence would be to drive the price of mercury up, and make the smuggling of it economically attractive. Mercury use would continue, but would become clandestine and therefore even more difficult to control. There are promising possibilities to improve the mining process, such as the likelihood that simple attachment to sluices could reduce mercury spillages at no cost to the miners, perhaps even increasing gold production at the same time. For such work to be possible, in addition to the kind of large-scale monitoring necessary to establish the extent of mercury contamination and begin to remedy it, it will be necessary to work *with* garimpeiros, not against them. This will also require appropriate laboratory facilities within Amazonia, and the training of Amazonians. With the necessary technical infrastructure in place within the region, the problems described here can begin to be addressed.

Revegetation of Metalliferous Wastes and Land After Metal Mining

M. S. JOHNSON, J. A. COOKE, AND J. K. W. STEVENSON

1 Introduction

There has been a progressive worldwide increase in metalliferous mining in recent years underpinned by social, economic, and technological demand.[1] The ever-increasing need for metals, and the ability of modern mining and processing methods to develop low grade ore-bodies economically, has placed increased strain on the environment at a time when the demands for high environmental standards are also increasing.[2] Since mining is, by its very nature, a destructive industry, attention has become focused on ways in which the environmental impact may be reduced or rendered temporary in nature.

An effective decommissioning plan for eventual mine closure which provides for the reclamation of disused workings, waste-rock dumps, and tailings impoundments, is an increasingly important factor when decisions are being taken as to whether to grant planning consent for a new mine.[3] This concern is underpinned by the legacy of past mining in many parts of the world. Continuing environmental damage arising from polluted waters and dispersal of contaminated solid waste is a feature of old mines in North America, Australia, Europe, and elsewhere.[4,5] It is, therefore, becoming standard practice for reclamation measures to be considered as an integral part of mine planning and operations, even to the extent that financial provisions are made during the operational life of a mine to effect such reclamation measures upon closure.[1] The higher environmental profile attached to modern mining is linked not only to social acceptability but also to legal requirements in many countries. This requires attention to be paid to the prevention of environmental damage from mining operations, waste production, and site closure.

[1] United Nations, 'Environmental Aspects of Selected Non-ferrous Metals Ore Mining', Technical Report Series No. 5, United Nations Environment Programme/Industry and Environment Activity Centre, UNEP, Paris, 1991, p. 116.
[2] United Nations, 'Mining and the Environment: The Berlin Guidelines', United Nations Department for Technical Co-operation and German Foundation for International Development, Mining Journal Books, London, 1992, p. 180.
[3] Ontario Ministry of Mines and Northern Development, 'Rehabilitation of Mines: Guidelines for Proponents', Ministry of Mines, Sudbury, Ontario, Canada, 1991, p. 137.
[4] G. M. Ritcey, 'Tailings Management: Problems and Solutions in the Mining Industry', Elsevier, New York, 1989, p. 970.
[5] N. J. Coppin, M. G. Staff, and M. S. Johnson, in 'Minerals, Metals and the Environment', Institute of Mining and Metallurgy, London, 1992, p. 104.

2   Constraints Upon Revegetation

Although chemical and physical techniques exist for dust control and stabilization against water erosion,[4] such objectives can only be realistically achieved in the long-term by the use of vegetation as a basis for landscaping, stabilization, and pollution control. However, it is widely recognized that wastes from metalliferous mines, especially acidic ones, are very difficult materials upon which to establish vegetation.[6]

The reasons that metal mine wastes present difficulties for plant growth derive from a combination of their physical, chemical, and biological properties. The major physical constraint of mine tailings is their small and uniform particle size distribution which is dominated by material of silt and clay dimensions. Physical properties vary both horizontally and vertically within any tailings impoundment due to stratification during deposition, and also to the particle sizing objectives set for optimum recovery of metals in the prior ore milling and flotation process. Tailings also possess unfavourable porosity, aeration, water infiltration, and percolation properties, along with a high bulk density and an absence of structural aggregates. The result is that water and wind erosion of unprotected disused tailings surfaces is a common hazard.[7]

In the case of old mines, often long abandoned, the processing technology deployed to recover the target metals was elementary and inefficient as compared with contemporary magnetic, gravity, and flotation processes. The consequence is that waste materials representing past eras of mining contain high residual quantities of toxic metals, often in excess of 1% by weight of elements such as lead, zinc, and/or copper (Table 1). These high metal values are frequently accompanied by elevated levels of non-target elements such as arsenic and cadmium from the original crude ore.[8]

Some metals (*e.g.* copper, zinc) are essential trace elements at low concentrations but toxic to plants at high levels. It is not possible to state specific concentrations for normal metabolism as opposed to toxicity as the threshold varies with other soil variables and the species concerned. Also, some combinations of metals act antagonistically or even synergistically in solution. Nickel and zinc, copper and zinc, and copper and cadmium are more toxic than their individual toxicities would suggest.[9] The presence of phosphate or calcium can reduce the toxicity of lead, zinc, and copper through precipitation and ion competition reactions. Other non-essential elements (*e.g.* lead, mercury) are less toxic to vegetation but hazardous to livestock that may graze vegetation which has accumulated such metals either via the roots or as surface dusts.[10]

The suite of problems described is compounded, in many instances, by what is

[6] M. S. Johnson and A. M. Mortimer, in 'Environmental Aspects of Metalliferous Mining: A Select Bibliography', Technical Communications, Letchworth, UK, 1987, p. 212.

[7] J. M. Ringe and D. H. Graves, *Reclam. Reveg. Res.*, 1987, **6**, 121.

[8] N. A. Williamson, M. S. Johnson, and A. D. Bradshaw, in 'Mine Waste Rehabilitation: the Establishment of Vegetation on Metal Mine Waste', Mining Journal Books, London, 1982, p. 103.

[9] C. G. Down and J. Stocks, 'Environmental Impact of Mining', Applied Science, London, 1977, p. 371.

[10] Department of Environment, 'Notes on the Restoration and Aftercare of Metalliferous Mine Sites for Pasture and Grazing', ICRCL Guidance Note 70/90, Department of Environment, UK Government, London, 1990, p. 15.

Table 1 Copper, lead, and zinc in spoil from abandoned mines in Britain*

Mining region	Counties	Number of sites surveyed	Principal base metals		
			Cu	Pb	Zn
S.W. England	Devon and Cornwall	16	65–6140	48–2070	26–1090
W. and N.W. England	Shropshire and Cheshire	12	15–7260	840–26 000	980–21 000
N. Pennines	N. Yorkshire and Durham	8	20–140	605–13 000	470–28 000
S. Pennines	Derbyshire	17	23–97	10 800–76 500	12 700–42 000
Lake District	Cumbria	7	77–3800	2070–7630	4690–7370
Mid-Wales	Powys and Dyfed	10	67–195	1670–54 000	475–8000
N. Wales	Clwyd and Gwynedd	19	30–5750	6400–76 000	11 300–12 700
S. Scotland	Dumfries and Galloway	6	125–657	4730–28 300	1600–31 400
Normal agricultural soil			2–100	2–200	10–300

*All values in $mg\,kg^{-1}$ air dried substrate

probably the most intransigent revegetation problems faced by the mining industry. The presence of significant quantities of iron pyrites (FeS_2), which may not be removed during ore beneficiation, often leads to very acid waste as the mineral is comminuted during processing and then undergoes weathering to generate sulfuric acid.[11] Pyrite-bearing wastes disposed of at neutral or slightly alkaline pH can degrade within months or years to produce extreme acidity.[12] Initial chemical processes are probably the result of natural weathering but this oxidation and hydrolysis is then assisted by ferrous ion oxidizing bacteria, *Thiobacillus ferrooxidans*, which thrive at pH 1.5–3.0.[13,14]

The rate of oxidation of pyrite is influenced by the surface area of the material available for weathering.[15] Other factors also influence the rate of production of acid, namely the native carbonate content of the material, and the size, morphology, and type of the pyrite present. If the wastes contain only small quantities of native carbonates, acid regeneration may exceed the neutralizing potential, resulting in a significant decline in pH. The pH values of mine tailings and waste-rock range from below 2 to more than 8, depending on the gangue material, the pyrite content, and, in the case of tailings, the chemicals added in the mill. Gold mining wastes in South Africa have pH values as low as 1.5, and

[11] M. U. Ahmed, in 'Extraction of Minerals and Energy: Today's Dilemmas', ed. R. A. Deju, Ann Arbor Science, Ann Arbor, Michigan, 1974, p. 49.

[12] A. C. Hartley, *Aust. J. Soil Res.*, 1979, **17**, 355.

[13] N. V. Blesing, J. A. Lackey, and A. H. Spry, in 'Minerals and the Environment', ed. M. J. Jones, Institute of Mining and Metallurgy, London, 1975, p. 341.

[14] J. R. Hawley, 'The Problem of Acid Mine in the Province of Ontario', Ministry of the Environment, Ontario, 1977, p. 32.

[15] M. Kalin, Proceedings of the 4th Annual General Meeting of Biominet, Sudbury, Canada, Special Publication, CANMET No. SP, 87–10, 1988.

uranium tailings in Colorado have been reported with pH values of 8.0.[16]

A further consistent feature of metalliferous waste, whether rock or tailings, is the low concentration of essential plant nutrients. Nitrogen levels are invariably inadequate for plant growth, phosphorus levels are generally very low, and deficiencies of potassium, calcium, and magnesium also may occur.[17] In warmer climates than temperate Britain, salinity of tailings can also prove troublesome.[18] It results from: (1) the interaction of the products of pyrite weathering with native carbonates; (2) the concentration of naturally occurring salts in tailings due to recycling of water; (3) the additions made to tailings by mill personnel in order to adjust effluent pH; and (4) excessive evaporation from the surface. The resulting salinity levels can be high enough to prevent plant growth. Tailings generally are more complex chemically than waste-rock, due to the addition of reagents and pH modifiers during the metal extraction process.[19]

The basic combination of adverse physical and chemical characteristics produces an environment in mine wastes that is hostile to plants. This is accentuated by the absence of the organic fraction that comprises 3–5% of the surface horizons of most natural soils. Organic matter contributes to soil structure, provides a reservoir of essential macronutrients, and a resource for invertebrates and micro-organisms that support the decay processes which underpin energy and nutrient cycling. The sterile nature of mine wastes has to be rectified before adequate and sustainable growth of plants can be achieved.

3   Revegetation Objectives

Essentially, the objectives of vegetation establishment are: long-term stability of the land surface which ensures that there is no surface erosion by water or wind; reduction of leaching throughputs, lessening the amounts of potentially toxic elements released into local watercourses and to groundwaters; development of a vegetated landscape or ecosystem in harmony with the surrounding environment; and with some positive value in an aesthetic, productivity, or nature conservation context.

The first objective is achievable by a continuous vegetation cover, especially where the cover is at least 100% and of relatively low growth. With lower cover values erosion may begin to occur.[20] The degree to which the second objective is met by a vegetation cover depends on the ambient climate. Vegetation will intercept and return rainfall to the atmosphere by evapotranspiration. In temperate climates the amount intercepted and returned will be up to 50% of the total, in the wet tropics less than 25%, and in the dry tropics more than 75%. However, these values are also affected by the distribution of the rainfall and the

[16]   H. B. Peterson and R. F. Nielson, in 'Ecology and Reclamation of Devastated Land', Gordon and Breach, New York, 1973, Vol. 1, p. 15.

[17]   A. D. Bradshaw and M. S. Johnson, in 'Minerals, Metals and the Environment', Institute of Mining and Metallurgy, London, 1992, p. 481.

[18]   K. L. Ludeke and A. D. Day, *Trans. Soc. Min. Eng. AIME*, 1985, **278**, 1807.

[19]   L. C. Bell and M. Evans, *Reclam. Rev.*, 1980, **3**, 113.

[20]   A. D. Bradshaw and M. J. Chadwick, 'The Restoration of Land', Blackwell Scientific Publishers Ltd., Oxford, 1980, p. 297.

type of rain event; it is more difficult for vegetation to intercept and re-evaporate precipitation if it falls in the form of intermittent, heavy storms. The third, and ultimately most demanding, objective represents the successful integration of all the factors that can influence a revegetation scheme and as such it is the ultimate test of the approach or the philosophy of the programme in practice.

4   Philosophies of Revegetation

The approaches to revegetation can be described in terms of three different basic philosophies: (1) ameliorative, (2) adaptive, and (3) agricultural.

(1)   *The ameliorative approach* relies on achieving optimum conditions for plant growth by improving the physical and chemical nature of mine wastes using organic matter, fertilizer, and/or lime. Alternatively, mixing or cover materials may be deployed. The most suitable species available commercially are sown on to the wastes whose edaphic properties have been modified in accordance with the land use objectives and type of vegetation to be introduced. This approach is commonly used in preference to the adaptive because it is quicker, requires less forward planning, and is less labour-intensive.

(2)   *The adaptive approach* emphasizes selection of the most suitable species, sub-species, cultivars, and ecotypes to meet the rigours of the extreme conditions. In addition, but not necessarily, the mine wastes may be improved using amendments to achieve optimum establishment and long-term growth. This approach is simple but is constrained by the availability of suitable propagules in some areas and by the long lead-time in producing commercial seed from promising natural or artificially selected plant material.

(3)   *The agricultural and forestry approach* has been used directly on less toxic media such as ironstone and bauxite wastes, and on wastes covered over with deep layers of soil or overburden. Agricultural crops or livestock or woodland and/or scrub species are established using conventional or specialized techniques.

In practice, it is combinations of the above approaches based on site-specific considerations that produce the final revegetation strategy. An extension to the above philosophies, and their combination, is the 'ecological approach' which places emphasis on the importance of establishing biological processes such as nitrogen fixation, decomposition, nutrient cycling and retention, and important biotic interactions (*e.g.* pollination). It is these that indicate proper ecosystem functioning, which is as important as the careful selection of plant species in providing the primary vegetation structure.[21] Whether the reclamation goals are to restore the original natural ecosystem or to produce an acceptable alternative, ecological principles should underlie all good reclamation schemes.

[21] A. D. Bradshaw, in 'Restoration Ecology', ed. W. R. Jordan, M. E. Gilpin, and J. D. Aber, Cambridge University Press, Cambridge, 1990, p. 53.

5 Revegetation Techniques and Land Use

Identification and treatment of the problems preventing plant growth, coupled with careful selection of species and appropriate long-term management, is the basis of successful revegetation. Various techniques have been developed to suit particular waste problems, ranging from cultivation with conventional agricultural machinery followed by fertilization and direct seeding for innocuous wastes, to specialist procedures such as placement of a barrier layer or deep coverings of non-toxic material for very toxic sites. The range of specific revegetation options together with their limitations are outlined in Table 2, and are based upon the degree of toxicity, salinity, and acidity of the waste material or site.

Many factors have to be considered in the choice of plant materials, and their method of establishment, in particular the nature of the spoil, the prevailing climate, and the eventual land use (Table 3). Examples of other local factors to be considered, including pest and disease incidence and availability, are given in (Table 4).

6 Direct Seeding

Normal Species

Unfortunately, straightforward direct seeding with conventional species and fertilizers is often unsuccessful as a revegetation measure, at least on older mine tailings, because of the toxic residual levels of metals—often with an acidity problem as well. Under these circumstances grass and other seedlings persist for only a few weeks. However, it remains an attractive option in principle because direct seeding is much cheaper than any other method. In situations where the waste has little residual metal, or where the metal is not available to plants, normal species can be established directly with the assistance of fertilizer.[22] Because the long-term growth of vegetation depends on an adequate supply of nitrogen, legumes such as white clover (*Trifolium repens*) or birdsfoot trefoil (*Lotus corniculatus*) are an important component of the seed mixture, since they have the capacity to supply nitrogen by fixation of atmospheric sources.[8]

Metal Tolerance

A close examination of even the most toxic waste from old metal mines representing eras when processing technology was crude and inefficient, nearly always reveals a sparse natural vegetation cover. Sometimes this is limited to only a few plants but of a characteristic and narrow range of species. Supplied with fertilizer, these plants are apparently able to tolerate and indeed thrive under conditions where non-tolerant plant material dies in a short time (see Figure 1). It is now known that these natural colonizers are special metal-tolerant populations of normal species that have become genetically adapted to thrive on metal-contaminated sites.

Metal tolerance is a widely recognized phenomenon in higher plants and has

[22] M.S. Johnson, A.D. Bradshaw, and J.F. Handley, *Trans. Inst. Min. Metall.*, 1976, **81**, A32.

Table 2 Approaches to revegetation

Waste characteristics	Revegetation technique	Problems encountered
Low metal toxicity. No major acidity or alkalinity problems.	(1) *Amelioration and Direct Seeding with Agricultural or Amenity Grasses and Legumes* Apply lime if pH < 6. Add organic matter if physical amelioration required. Otherwise apply nutrients as granular compound fertilizers. Seed using agricultural or hydraulic techniques (*e.g.* $Pb/Zn/CaF_2$ tailings, Derbyshire, UK).[22]	Probable commitment to long-term maintenance. Grazing must be strictly monitored and controlled in some situations due to movement of toxic metals into vegetation.
Low metal toxicity and climatic limitations. No major acidity or alkalinity problems. Extremes of temperature, rainfall, *etc.*	(2) *Amelioration and Direct Seeding with Native species* Seed or transplant adapted native species using amelioration treatments (*e.g.* lime, fertilizer) where appropriate (*e.g.* Cu tailings, Arizona, USA).[23] Use thin layer of native soil as seed inoculum.	Irrigation often necessary during establishment in arid climates. Expertise required on the selection of native flora. Labour intensive.
Medium to high metal toxicity. High salinity.	(3) *Amelioration and Direct Seeding with Tolerant Ecotypes* Sow metal and/or salt and/or acid tolerant seed. Apply lime, fertilizer, and organic matter, as necessary, before seeding (*e.g.* Pb/Zn waste Wales, UK).[24]	Possible commitment to regular fertilizer applications. Relatively few species have evolved tolerant populations. Grazing inadvisable. Very few species are available commercially as tolerant varieties.
	(4) *Surface Treatment and Seeding with Agricultural or Amenity Grasses and Legumes* Amelioration with 10–50 cm of innocuous mineral waste (*e.g.* overburden). Apply lime, fertilizer and organic matter as necessary (*e.g.* Pb/Zn waste, Wales, UK).[25]	Regression will occur if shallow depths of amendment are applied or if upward movement of metals occurs. Availability and transport costs of surface amendments may be limiting.
Extreme toxicity. Very high toxic metal content. Intense salinity or acidity.	(5) *Barrier Layer* Surface treatment with 30–100 cm of innocuous barrier material (*e.g.* unmineralized rock) and surface covering with a suitable rooting medium (*e.g.* subsoil). Apply lime and fertilizer as necessary (*e.g.* Pb/Zn/Cu wastes, New South Wales, Australia).[26]	Susceptibility to drought according to the nature and depth of the surface covering. High cost and potential limitation of availability of barrier material. Integrity of barrier layer may be affected by root penetration.

[23] K. L. Ludeke, in 'Tailings Disposal Today', ed. C. L. Aplin and G. O. Argall, Miller Freeman, San Francisco, 1973, p. 606.

[24] R. A. H. Smith and A. D. Bradshaw, *J. Appl. Ecol.*, 1979, **16**, 595.

[25] M. S. Johnson, T. McNeilly, and P. D. Putwain, *Environ. Pollut.*, 1977, **12**, 261.

[26] B. Craze, *J. Soil Conserv. Serv.*, 1977, **33(2)**, 98.

Table 3 Primary considerations in selection of plants

Primary considerations	Plant types selected
Nature of Spoil	
Toxic metals at high concentrations	Metal-tolerant cultivars Natural invaders of mineralized outcrops
Toxic metals moving into herbage	Unpalatable species Spiny shrubs around site perimeter
Extreme acidity/alkalinity	Natural invaders of acidic or alkaline conditions
High levels of salts	Salt-tolerant species Natural invaders of salty areas
Drought conditions	Drought-tolerant species Certain metal-tolerant cultivars
Poor nutrient status	Legumes or other nitrogen-fixers Species that grow in nutrient poor areas
Climate	
Extreme cold with short growing season	Native or naturalized species Species that grow and develop rapidly
Arid or semi-arid	Native or naturalized species Transplants or cuttings of slow growing species
Temperate	Agricultural, forestry, or other commercial species depending on land use
Eventual Land Use	
For rapid stabilizing cover and high productivity	Agricultural species
For wildlife	Variety of native and naturalized species that provide seeds, fruits, palatable herbage, nesting sites, *etc.*
For aboriginal or tribal use	Native species Timber, medicinal, or food crops Species that regenerate after practices such as burning of forests
For amenity and recreation	Wear-tolerant cultivars as developed for sportsground turf Low productivity

Table 4 Other considerations in selection of plants

(1) *Insect resistance*

River red gum (*Eucalyptus camaldulensis*) has proven to be a most promising reclamation tree worldwide. Its performance at Groote Eylandt manganese mine in Northern Territory, Australia, has also been promising but its lack of resistance to termites could impose considerable limitations on its timber value, a prime consideration in this reclamation scheme.[27]

(2) *Disease resistance*

Disease resistance is becoming an important factor in selection of species for reclamation of bauxite-mined lands in the Jarrah forests of the Darling Range, Australia, where *Phytophthora cinnamomi*, a soil borne fungal pathogen, is causing large-scale dieback of vegetation.[28]

(3) *Intrusion by man*

Thickets of spiny shrubs such as Arnot Bristly Locust (*Robinia fertilis*), an acid-tolerant, nitrogen-fixing shrub, may be effectively positioned around perimeters of open pits and radioactive or hazardous wastes to restrict public access.

(4) *Landscape planting*

Trees with rapid growth such as Black Locust (*Robinia pseudoacacia*) can be effective in visual screen plantations for tailings ponds, waste heaps, and mine buildings.

(5) *Growth habit*

Ideally, material should be easily propagated, quick to establish, be mat-forming with fibrous root systems or rhizomes. Selection of deep rooted plants such as alfalfas and trefoils may be of value in breaking up a compacted soil. Herbaceous or perennial plants should be favoured rather than annuals, which suffer problems with re-establishment.

(6) *Competition*

Species should be chosen that grow favourably with other components of the mixture; for example, establishment of young trees in a lush ground cover vegetation may be adversely affected by competition.

(7) *Availability*

If possible, species should be selected that are available commercially. If companies can foresee requirements, and order from reputable commercial nurseries giving at least a year's notice, unusual requirements can often be met. If full reinstatement of a diverse native flora is to be carried out, then the company should consider establishing its own nursery facilities.

---

[27] P. Langkamp and M. Dalling, *BHP J.*, 1977, **1**, 42.

[28] J. R. Bartle and S. R. Shea, Proceedings of the Australian Mining Industry Council (1979), Environmental Workshop, Bunbury, Australia, 1979.

Figure 1 Growth of lead/zinc tolerant and non-tolerant populations on lead/zinc mine waste after seven months growth

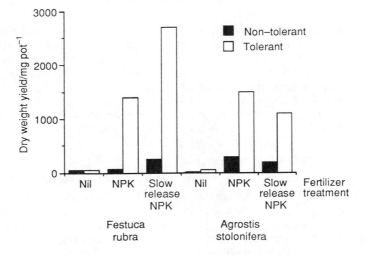

been reported in respect of cadmium,[29] arsenic,[30] and nickel[31] as well as in the more widely publicized cases of copper, lead, and zinc.[32] Recognizing the revegetation potential of these plants, a breeding programme was initiated in the 1970s in order to produce commercial cultivars of certain grasses bearing the same genetic characteristics as the 'natural' plant material. As a result of this, three cultivars of temperate grasses are now available commercially enabling direct seeding of toxic areas. These three, and their associated tolerances, are: *Festuca rubra* cv. 'Merlin' (lead–zinc), *Agrostis capillaris* cv. 'Parys' (copper), and *Agrostis capillaris* cv. 'Goginan' (lead–zinc). The value of these cultivars for seeding unstable and toxic tips is considerable since revegetation can be achieved by direct sowing and treatment with simple inorganic fertilizers.[24] They are particularly useful for revegetation of older sites, though may prove to be less essential on newer waste materials as the improved mineral processing technologies of today leave less residual metal in the tailings. Under these circumstances, normal plant material is able to thrive (see Figure 2).

Overall, revegetation costs are low using the tolerance route but there are limitations. For example, the tolerance mechanism is specific and though cross-tolerance exists between related metals,[32] this is at a low level so a particular cultivar is unlikely to thrive on mine waste containing toxic quantities of metals other than those to which it has evolved an adaptation. Furthermore, grazing of the sward for agricultural purposes is not usually possible and the recreational or trampling resistance of the ground cover comprising tolerant plant material is low. Also, long-term fertilizer input is usually required and from an ecological viewpoint it is important that such schemes using metal-tolerant plants can lead eventually to the site becoming available for colonization by other tolerant but also non-tolerant species.

[29] P.J. Coughtrey and M.H. Martin, *New Phytol.*, 1978, **81**, 147.

[30] J.R.C. Hill, *Trans. Inst. Min. Metall.*, 1977, **86A**, 98.

[31] R.R. Brooks, J. Lee, R.D. Reeves, and T. Jaffre, *J. Geochem. Explor.*, 1977, **7**, 49.

[32] A. Baker, *New Phytol.*, 1987, **106**, 93.

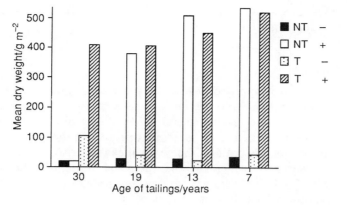

Figure 2 Growth of lead/zinc tolerant (T) and non-tolerant (NT) *Festuca rubra* with (+) and without (−) NPK fertilizer on Pb/Zn/CaF$_2$ tailings of different ages, in Derbyshire, UK

Considering the usefulness of tolerant temperate grasses in revegetation work it is surprising that equivalent commercial breeding programmes have not so far taken place for some of the more promising tropical species, including *Chloris gayana* and *Eragrostis curvula*. These could possibly be developed for mine waste revegetation.

7   Surface Improvement and Covering Systems

Principles

The selection of methods for improving growing conditions should be based on: (1) alleviating toxicity and acidity; (2) augmenting supplies of essential plant nutrients; (3) improving the physical properties; and (4) achieving maximum benefit from the materials available on site or nearby. Fertilizer application to tailings is always necessary and use of organic matter is advisable if it can be obtained locally. Correction of acidity or alkalinity not only enables a wider range of plants to be established, but also alleviates metal toxicity and increases availability of nutrients for plants.

The principle behind addition of materials to tailings or covering over of the surface is to **dilute** or **avoid** toxicity problems rathern than **counter** them by direct seeding of tolerant populations. The covering of mine waste to isolate it from the establishing vegetation is a common approach to reclamation and can succeed if a suitable depth of material can be introduced into which the chosen vegetation can root and develop satisfactorily. Usually the cover material is topsoil, subsoil, or overburden. However, it is rarely feasible, for economic reasons, to provide depths of cover greater than 300 mm, and in the case of some modern tailings the load-bearing capacity precludes the use of most forms of civil engineering equipment required to apply the cover. On wastes of low to medium toxicity, covering layers can provide a cheap method of improvement, whilst very toxic materials require barrier layers or isolating materials between the waste and growing medium to reduce upward movement of metals. In some cases, especially where toxicity is marginal, simple dilution of the waste with innocuous material may suffice.

Dilution

The simplest approach to revegetation using amendments is to incorporate suitable material into the mine waste surface on the principle of diluting the influence of the residual metal values below phytotoxic thresholds. Organic matter, in particular, is used in this connection because it has important beneficial effects both on the physical characteristics and the nutrient status of mine wastes.[33] It increases the water and nutrient-holding capacity, improves surface stability, aeration, and water penetration by alteration of the soil structure, whilst decreasing surface run-off, and improving the seed bed. In addition, heavy metals can be temporarily complexed or chelated by organic material which binds them in an innocuous form until natural decay of the organic matrix causes remobilization. Organic materials also provide a source of micro-organisms. Amendments such as farmyard or poultry manure or sewage sludge, are usually incorporated into the waste surface to 150 mm depth by discing. The aim is to achieve about 3–6% organic matter content, which is the level expected in a normal soil.[34]

Modern high-analysis, compound fertilizers are used in association with the dilution and cover approaches to revegetation. They are formulated from compatible chemicals, and are easy and clean to handle whilst occupying less storage and transport space than the more bulky organic sources of nutrients. Compound fertilizers are available to cover most needs, *e.g.* NPK fertilizer 17.17.17, which supplies 17% N, 17% P_2O_5 (7.5% P), and 17% K_2O (14% K) by weight. Slow-release commercial fertilizers, which release their nutrients in a time-graded pattern over months and even years, are more expensive but can produce good results and reduce labour costs.[20]

Fluorspar–lead–zinc mine tailings in the Peak District National Park in the UK, represents a successful example of the dilution approach to reclamation. Residual levels of metals and fluorides proved not to be toxic and a hydroseeding technique was used to establish a commercial grass–legume seed mix, with the use of air-dried, digested sewage sludge and phosphate fertilizer applied directly to the dewatered tailings surface.[22] The establishment of the grassland together with subsequent planting of trees and shrubs has been very successful and led to extensive colonization by wildlife.

Surface leaching, using overhead sprinkler systems as the basis for dilution of acidity, has been used successfully along the Witwatersrand near Johannesburg in South Africa for controlling acid–sulfate levels in the surface layers of gold mine tailings. Regular mist-spraying regulates the acidity in surface layers at a sufficiently low level to permit the establishment of grasses (*e.g. Chloris gayana, Cynodon dactylon, Eragrostis curvula*) and legumes (*e.g. Medicago sativa*) after treatment of the tailings with lime and fertilizer.[35] The technique, though

[33] S.J. Stokowski, Jr., Proceedings of the Annual Meeting of the Society of Mining, Phoenix, Arizona, 1988.

[34] G.W. Cooke, 'The Control of Soil Fertility', Crosby Lockwood, London, 1967, p. 526.

successful, is expensive in equipment and labour, reflecting the intransigence of highly pyritic metalliferous wastes to long-term reclamation.

Simple Coverings

Some form of soil or surface cover has been extensively used in past revegetation schemes to avoid toxicity and to improve the texture and stability of waste surfaces so that vegetation can be established. Many of these schemes have worked well but failures have also resulted because of: (1) lack of penetration of roots into the underlying material leading to poor binding at the soil/waste interface; (2) contamination of the covering through upward migration and accumulation of toxic metals, salts, and acidity; or (3) penetration of plant roots into toxic material beneath, with subsequent regression of the vegetation. Experience suggests that the minimum depth of such surface coverings should be 300 mm.[36] With less, erosion of the covering is a real risk.

Coastal dune mining for heavy minerals at Richards Bay on the east coast of S. Africa illustrates the possibilities of reinstating the former vegetation using a shallow layer of topsoil directly on to non-toxic tailings. Topsoil, recovered in advance of the opencast dredgers, is returned to the reformed dunes because it contains a viable seed bank. After 8–12 years, 9 m high *Acacia karoo* woodland has established, which in most respects is comparable to natural stands of this age. There is evidence that normal succession will lead eventually to dune forest typical of the area.[37]

Barrier or Isolating Layers

The use of barrier layers, though less common than the simpler methods on the grounds of the higher cost, is nowadays popular for wastes that present a particular hazard to local communities through toxic metal pollution, and also where there is a pressing need to develop a specific land use (*e.g.* sports fields, grazing land), or where vegetative stabilization cannot otherwise be considered due to extremes of toxicity and acidity. If simple covering layers such as soil are used on toxic waste, then even with deep layers (> 300 mm) upward migration of contaminants may in time cause regression of vegetation. In these cases it is necessary to use barrier layers of material designed to inhibit the upward movement of solutes.

The main requirement is that the barrier layer should disrupt the capillary water columns established within the waste. Usually this means at least a 300 mm deep layer of a coarse textured material such as screened gravel, with no fines,[38] rock waste, or coarse non-toxic mine spoil. A column study in Canada, investigating the use of a barrier material, showed that a 50 mm layer of coarse

[35] Chamber of Mines, 'Handbook of Guidelines for Environmental Protection', Chamber of Mines of S. Africa, 1979, p. 5.

[36] M. A. Norem and A. D. Day, *Reclam. Reveg. Res.*, 1985, **4**, 83.

[37] P. Camp and P. J. Weisser, in 'Dune Forest Dynamics in Relation to Land-use Practices', ed. D. A. Everard and G. P. von Maltitz, Foundation for Development, S. Africa, 1991, p. 106.

[38] S. Ames, Proceedings of the 3rd Annual British Columbia Mine Reclamation Symposium, Vernon, British Columbia, Canada, 1979.

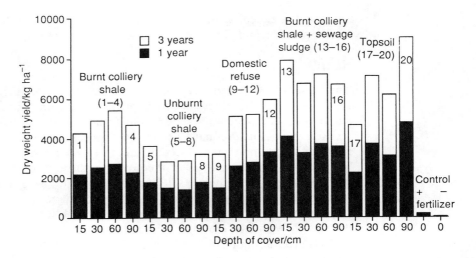

Figure 3 Dry weight yields of a grass–legume mixture on different types and depths of cover material on Pb/Zn waste after one and three years growth at Y Fan mine, Powys, West Wales

gravel placed between acidic iron tailings and a 450 mm layer of overburden, was effective in preventing the upward movement of salts, acids, and toxic metals.[38] On a field scale, however, there would be serious technical problems in spreading a 50 mm layer; 300 mm would be a more practical minimum as regards the use of large machinery. In practice therefore, a 300 mm layer of porous, coarse material covered with a minimum 300 mm depth of rooting medium is usually recommended. In order to be effective, the barrier layer must be free of soil or organic matter and allow free lateral and downward drainage of infiltrating water. The rooting medium allows establishment and growth of stabilizing vegetation and storage of soil moisture. In warm climates, adequate storage of moisture cannot be achieved without deep surface layers.

Faced with the problems of achieving a permanent vegetation cover on as shallow a layer of imported material as possible, extensive trials were undertaken at Y Fan lead–zinc mine in west Wales between 1975 and 1982 (Figure 3). The objective was to develop treatment systems that would prevent root accumulation of toxic metals, eliminate vertical migration of soluble metal salts and provide a vegetation cover that would survive, independently of regular maintenance, because of the high management costs the latter incurs. The field experiments at Y Fan compared plant growth on various inorganic waste materials placed over highly toxic lead–zinc wastes. Figure 3 shows that burnt colliery spoil with sewage sludge was effective in achieving good plant growth and it also prevented movement of lead and zinc, in fact much better than more usual covering materials.[25] The colliery spoil in question was non-toxic, non-pyritic, and of a suitable texture to disrupt water columns in the tailings.

A covering approach has been used at the copper tailings dam of the old Avoca mine in Co. Wicklow, Republic of Ireland. This 30 ha dam was subject to serious erosion problems before the enactment of a revegetation scheme in 1984. The erosion problem was so serious that direct seeding using tolerant seed presented too great a risk in view of the relative slowness of sward establishment by this method. Accordingly, a two-layered cover approach was adopted in which a layer of shale, 200–300 mm deep, was placed on the tailings surface to isolate the material and then overlaid with a skim of 75–100 mm of topsoil and subsoil to provide the supportive medium for cover vegetation. The surface was then treated with conventional limestone and fertilizers before being sown with a traditional agricultural seed mixture.

With careful management, the results were outstanding in the first two years. A crop of hay was taken from the reclaimed surface in the summer following the year of reclamation. The quality of the product was such that there were no constraints upon feeding the product to livestock. In recent years, the management of the site has changed and the sward has been permitted, quite deliberately, to deteriorate and become invaded by other species. The grass surface now supports gorse (*Ulex europaeus*) and broom (*Sarothamnus scoparius*), plus a wide range of herbaceous species, and is now carrying a significant complement of wildlife. In particular, it provides excellent cover for pheasant which have been bred for decades in local woodlands.

A long-standing example of revegetation based on a combination of the use of a shallow covering layer and tolerant plants together is that of Parc Mine, Nr. Llanwrst in north Wales. This old, abandoned lead–zinc mine had a tailings dam of 6 ha that, until 1977, was undergoing severe erosion caused by the steep, angular shape of the dam and the absence of a proper retaining wall. The dam was re-contoured and 200 mm cover of quarry shale placed as a single layer, before seeding with a mixture based on *F. rubra* cv. 'Merlin' with accompanying white clover (*Trifolium repens*). The advantage of including this tolerant fescue in the final product was that the shallow layer of cover was adequate for the clover but the fescue roots penetrated into the tailing beneath keying the cover to the tailings and increasing the physical stability and thus the persistence of the sward. The specification proved successful and an effective cover has been maintained to the present day.

The same broad style of revegetation has been used in the Mediterranean region, particularly in Spain, for decommissioning tailings dams where wind erosion and the residual copper levels are just too high to enable the simpler direct seeding systems described earlier to succeed. In this instance infertile but weathered surface rock and shale was stripped from the base slopes of pine woodland adjacent to the tailings area, and placed upon the tailings surface as a shallow, 200 mm, layer. Trials showed that native buried seed from the natural seed bank (particularly *Pinus pinaster* and *P. pinea*) provided sufficient viable propagules for re-establishment of woodland without any further effort or expenditure apart from the use of limestone and conventional fertilizers as aids to establishment (Figure 4).

This approach, simple covering with a shallow layer of local seed-rich soil and application of NPK fertilizer at 400 kg ha^{-1}, has been outstandingly successful on a large scale tailings facility of 60 ha. In northern Spain, where the climate is sub-Mediterranean, the cover approach has been employed successfully using non-toxic mine overburden. Following trials, direct seedings of tree and shrub seed on to a 500 mm layer of overburden was undertaken on two areas of tailings at a disused copper mine. Slow-release NPK fertilizers were used together with a

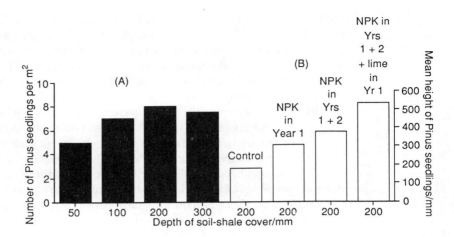

Figure 4 (A) Establishment of *Pinus* seedlings from buried seed banks of soil/shale cover in relation to depth of cover applied. (B) Mean height of *Pinus* seedlings 2 years after seeding and in relation to fertilizer treatment

seed mixture dominated by *Eucalyptus globulus* and *E. camaldulensis*, along with *Acacia retinoides*, *Pinus* spp., and *Betula* spp. as nurse species. The dual reclamation objectives of erosion control and amenity improvement were met less than three years after seeding.

8 Management and Aftercare

Where toxic wastes are reclaimed, regression of a well-established sward can occur. Regression may be due to one or more of the following: (1) weathering of pyritic wastes producing acidity, which in turn alters the availability of plant nutrients and toxic metals; (2) gradual decomposition of organic amendments releasing metals previously held in stable organic complexes; (3) depletion of nutrients required for growth; (4) extreme weather conditions; or (5) upward migration of acidity, heavy metals, or salts into the surface layers of amendment.[39] Long-term management should therefore be considered as an integral part of any reclamation scheme and should be planned at an early stage.

The programme of long-term management will depend ultimately upon the species sown and the land use objective. Refertilization and/or management of a legume component, cutting/grazing, pruning and tying of trees, and fencing maintenance may all figure as components of a management programme for a reclamation site. It is necessary to build up soil fertility, in particular with respect to nitrogen, and then to maintain plant growth through the establishment of leguminous species and the decomposition of organic matter. Legumes are encouraged by maintaining a suitable pH (>5.5) and by application of phosphate fertilizer. In the period directly following reclamation, re-treatment with inorganic fertilizers is necessary, and on acidic wastes maintenance dressings of lime may also be required. Fertilizer applications are based on assessments of the extent of ground cover, colour of the vegetation, seeding potential, productivity, and the results of nutrient analyses and possibly small-scale field

[39] J. R. Harris and A. I. M. Ritchie, *Environ. Geochem. Health*, 1987, **9(2)**, 27.

experiments. Generally, fertilizer inputs are necessary for 4–6 years. These annual applications should be in the order of 35–70 kg ha^{-1} each of N, P_2O_5, and K_2O,[20] depending on vegetation type and land use objective, *etc.*

Grasslands, especially more productive swards comprising forage species, require cutting, grazing, or burning to prevent their gradual deterioration. These practices thicken the sward by encouraging tillering. The timing and frequency of cutting and grazing influence the species that establish, those that persist over time, and the natural seeding that takes place. One of the main problems of aftercare on metal mine wastes results from mobilization of toxic metals, leading to uptake by plants, and potential toxicity problems to grazing livestock. Regular monitoring of metals in vegetation is necessary, and prompt attention to any increase in metal content should enable potential problems to be avoided.

9 Conclusions

Standards of environmental care within the mining industry have increased greatly in the last decade, and derive from the recognition of the socio–environmental demands of society, as well as the need to comply with statutes and regulations that are extending worldwide in order to improve performance and accountability. In the last two decades, techniques have been developed that enable revegetation of metalliferous mine wastes to be undertaken to a necessarily high standard, thus improving the prospects for creating a new, substitute landscape that, whilst not the restoration of the original, may be viewed as a satisfactory and sometimes even more valuable replacement.

The essential ingredients of success are commitment, careful development of a specification, specific objectives, practical implementation methods—sometimes deploying innovative ideas and equipment—and a properly formulated management and aftercare plan (Figure 5). Different endpoints are achievable according to the waste, location, and climate. A wide range of land uses is available, including various grasslands, woodland, and wildlife habitat. On occasions it is even possible to regard some form of crop or livestock production as a part of the long-term revegetation plan. However, there remain some doubts as to the sustainability of certain specifications in the long-term. The so-called 'walk-away' option has been promoted at some sites but remains an elusive target. What is certain, however, despite the challenges that remain, is that with modern reclamation techniques it is no longer appropriate to argue against mining in its totality based upon the inevitability of irreversible damage to the landscape, loss of amenity, or nature conservation values.

Figure 5 Flow chart of the stages in achieving successful revegetation of mine waste or tailings

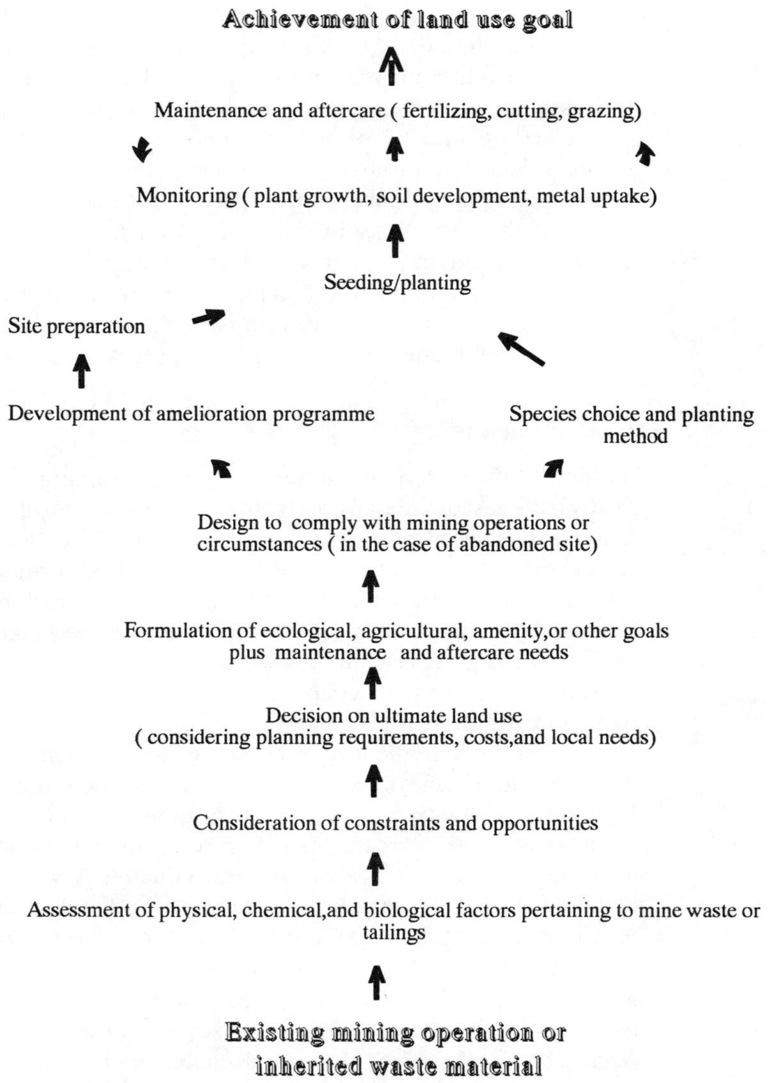

Achievement of land use goal

⬆

Maintenance and aftercare (fertilizing, cutting, grazing)

⬇ ⬆ ⬆

Monitoring (plant growth, soil development, metal uptake)

⬆

Seeding/planting

Site preparation ➡

⬆

Development of amelioration programme

Species choice and planting method

Design to comply with mining operations or circumstances (in the case of abandoned site)

⬆

Formulation of ecological, agricultural, amenity,or other goals plus maintenance and aftercare needs

⬆

Decision on ultimate land use (considering planning requirements, costs,and local needs)

⬆

Consideration of constraints and opportunities

⬆

Assessment of physical, chemical,and biological factors pertaining to mine waste or tailings

⬆

Existing mining operation or inherited waste material

Vegetative Remediation at Superfund Sites

G. M. PIERZYNSKI, J. L. SCHNOOR, M. K. BANKS,
J. C. TRACY, L. A. LICHT, AND L. E. ERICKSON

1 Introduction

Non-ferrous metal mining activities across the world have produced a variety of environmental problems. Three types of contamination created by large-scale metal extraction have been identified.[1] Waste-rock, tailings, and slag are primary contaminants. Secondary contamination occurs in groundwater beneath open pits and ponds, sediments in river channels and reservoirs, floodplain soils impacted by contaminated sediment, and soil affected by smelter emissions. River sediments reworked from floodplains and groundwater from contaminated reservoir sediments were identified as tertiary contaminants. In the United States, many of these contaminated sites have been classified as Superfund sites, which dictates that some remedial action be taken in the future.

The metals or metalloids most commonly found at Superfund sites are arsenic (As), cadmium (Cd), chromium (Cr), copper (Cu), lead (Pb), mercury (Hg), nickel (Ni), selenium (Se), silver (Ag), and zinc (Zn).[2] Most of the discussion will focus on As, Cd, Pb, and Zn as elements of concern for two Superfund sites: the Whitewood Creek in the Black Hills of western South Dakota and the Galena site located in southeastern Kansas. Contamination of surface water and groundwater with As and Cd from over one hundred years of gold mining activity is the principal concern at the Whitewood Creek site. The Galena site is located in the Tri-State mining region (southeast Kansas, southwest Missouri, and northeast Oklahoma), where Pb and Zn sulfide ores were mined and smelted extensively from the mid-1800s to approximately the 1950s. Pb and Zn contaminated mine spoils, soils, groundwater, and surface water are extensive problems in the Tri-State Region.

There are two primary reasons for concern over elevated concentrations of trace elements in waters, soils, or mine spoils. First, elevated human and animal exposure to the metals can occur through food chain transfer, ingestion of wind-blown dusts, or direct ingestion of soil. Persons living downwind of an old smelter site in the Tri-State region could consume at least 50% more Pb and Cd by eating some of their home-produced food items than by eating comparable

---

[1] J. N. Moore and S. N. Luoma, *Environ. Sci. Technol.*, 1990, **24**, 1278–1285.

[2] J. E. McLean and B. E. Bledsoe, 'Behavior of Metals in Soils', EPA Ground Water Issue, EPA/540/S-92/018, US Environmental Protection Agency, Washington, DC, 1992.

Table 1 Remediation options for metal-contaminated sites

Method	Comments
Excavation followed by:	
Solidification	Addition of cementing agent to produce a hardened, non-porous, non-leachable material.
Vitrification	Heating to produce a glass-like, non-porous, non-leachable material.
Washing	Chelate or acid extraction.
Leaching	Pile or batch leaching with chelates or acids.
Particle size segregation	Selective removal of finer particle sizes (*e.g.* clay) that have the highest metal concentrations.
In situ	
Solidification	As described above.
Vitrification	As described above.
Encapsulation	Cover site with impermeable layer.
Attenuation	Dilution with uncontaminated material.
Volatilization	Promote formation of volatile methylated species (Se, As, Hg).
Vegetative	Promote vegetative growth by providing proper fertility and water availability, reducing metal bioavailability, and/or using metal-tolerant plant species.

items purchased in a control area.[3] Epidemiological studies have shown a significantly higher prevalence of chronic kidney disease, heart disease, skin cancer, and anemia in persons living for more than 5 years in Galena, KS, than in the populations of two nearby control towns.[4] Inhalation of As has been associated with lung cancer, and ingestion of As is judged to cause skin cancer.[5] The second reason for concern relates to the phytotoxic potential of the metals, which can limit biomass production.[6,7] This inhibition of plant growth can have direct negative effects, such as a limitation of crop yields. The effects also can be indirect. For example, the lack of vegetative cover probably will result in enhanced wind and water erosion, which further disperses the contaminants and increases the likelihood of human exposure via wind-blown dusts.

Numerous remediation options exist for metal-contaminated sites, as shown in Table 1. An excellent description of some experimental methods has been published.[8] The methods requiring excavation have a significant drawback given that the volume of material to be treated can be quite large. For example,

[3] J. V. Lagerwerff and D. L. Brower, in 'Trace Substances in Environmental Health, Vol. 8, ed. D. D. Hemphill, University of Missouri, Columbia, MO, 1974.

[4] J. S. Neuberger, M. Mulhall, M. C. Pomatto, J. Sheverbush, and R. S. Hassanein, *Sci. Total Environ.*, 1990, **94**, 261–272.

[5] D. W. North, *Environ. Geochem. Health*, 1992, **14**, 59–62.

[6] S. B. Bradley and J. J. Cox, *Sci. Total Environ.*, 1986, **50**, 103–128.

[7] G. M. Pierzynski and A. P. Schwab, *J. Environ. Qual.*, 1993, **22**, 247–254.

[8] United States Environmental Protection Agency, 'The Superfund Innovative Technology Evaluation Program: Technology Profiles', (ed. 5) EPA/S40/R-92/077. US Government Printing Office, Washington, DC, 1992.

Cherokee County, KS (which contains the Galena Superfund site) has numerous abandoned Pb and Zn mining and smelter sites. The soil survey for the county reports 1316 hectares of mine dump sites,[9] which have high Pb and Zn concentrations and would benefit from remediation. If only the top 300 mm of these areas were treated, this would involve approximately 4.8×10^6 Mg of material (1 Mg $\equiv$ 1 tonne). This is a conservative estimate, because most areas would require more than the top 300 mm be treated, and some areas that need remediation are not shown in the soil survey.

The beneficial effects of plants in remediation of soil and groundwater contaminated with hazardous organic compounds have been presented.[10] The vegetative remediation methods for metal contaminated sites, which are the focus of this paper, can utilize amendments that reduce metal bioavailability as well as metal-tolerant plant species with the goal of establishing a vegetative cover sufficiently dense to prevent wind and water erosion and that will remain viable for extended periods. The vegetation can be native or introduced grasses, forbs, or trees. The advantages of vegetative remediation include the minimization of wind and water erosion, lower cost as compared with other remediation options, improvement of aesthetics, no production of waste products, increases in soil organic C concentrations (binds metals, improves soil tilth, *etc.*), and the potential to serve as a temporary remediation until more suitable methods are funded or developed. In addition, modeling efforts suggest that vegetation, particularly trees, probably would reduce net percolation through the soil or mine spoil material and reduce the leaching potential of the metals. Disadvantages include the lack of data on the long-term viability of the vegetation, the possibility of producing metal-rich plants that could be consumed by wildlife or other animals, the lack of transpiration by the plants during certain periods of the year, and the possibility of transport of radionuclides or metals in mixed wastes due to excretion of soluble exudates by plant roots.

The goals of this article are to briefly describe the chemical and microbiological environment in mine spoils and contaminated soils, to describe several case studies where vegetation has been used in remediation of Superfund mine sites, and to present a generalized model that can aid in predicting the effects of vegetation on a contaminated site.

2 Chemical Aspects of Metal-contaminated Soils and Mine Spoils

Chemical characteristics such as total metal concentrations, pH, cation exchange capacity, plant nutrient concentrations, and organic C content in contaminated soils and mine spoils can vary considerably. English soils having less than 50% vegetative cover contained 1660 mg kg^{-1} Pb and 4230 mg kg^{-1} Zn in the surface 50 mm.[11] Soils with vegetation exhibiting heavy metal chlorosis had 323 mg kg^{-1} Pb and 676 mg kg^{-1} Zn in the top 50 mm.[6] Zn and Pb concentrations as high as

[9] Soil Survey Staff, 'Soil Survey of Cherokee County, Kansas', USDA Soil Conservation Service, US Government Printing Office, Washington, DC, 1985.

[10] J. F. Shimp, J. C. Tracy, L. C. Davis, E. Lee, W. Huang, and L. E. Erickson, *Crit. Rev. Environ. Sci. Technol.*, 1993, **23**, 41–77.

[11] M. S. Johnson and J. W. Eaton, *J. Environ. Qual.*, 1980, **9**, 175–179.

43 750 and 4500 mg kg^{-1}, respectively, have been reported for mine spoil material.[12] Gold mine tailing in South Dakota contained 917 mg kg^{-1} As.[13] In the United States, the Toxicity Characteristic Leaching Potential (TCLP) is used to classify materials as hazardous or not.[14] The procedure involves a single extraction with 0.1 M acetic acid in an effort to simulate leaching conditions that a waste might experience. If the concentrations of certain metals exceed some standard values, the material is classified as hazardous.

The pH of the contaminated soils or mine spoil materials can range from values as low as 2.0 to as high as 8.0. The very acid conditions typically are associated with the weathering of sulfide-bearing minerals. The alkaline conditions can be caused by the presence of a calcareous matrix. In terms of cation exchange capacities, plant nutrient concentrations, and organic C concentrations, one can consider contaminated soils and mine spoils as diluted soils. That is, these parameters will range from extremely low values (highly diluted) to those typical for soils (not diluted). Indeed, low fertility because of low cation exchange capacities and plant nutrient concentrations and low water holding capacities because of low organic C concentrations are as limiting as metal phytotoxicities in establishment of vegetation in mine spoil materials.

The behavior of metals in soils has been reviewed.[2] Most metals interact with the inorganic and organic matter that is present in the root–soil environment; potential pools or forms of metals include those dissolved in the soil solution, adsorbed to the vegetation's root system, adsorbed to insoluble organic matter, bonded to exchange sites on inorganic soil constituents, precipitated or coprecipitated as solids, and within the soil biomass. Generally, the total metal concentration in soil is a poor indicator of metal availability to plants. The concept of metal bioavailability, in the context of soils and mine spoils, refers to some sub-fraction of the total amount of a metal that best correlates to plant response. That response is typically measured in terms of biomass production or metal concentrations in plant tissue. Any of the pools or forms of metals described above can contribute to the bioavailable fraction. In practice, metal bioavailability is often operationally defined as that extracted with a particular extractant.

Metals present in the soil solution can be free metal ions, soluble complexes with organic or inorganic ligands, or associated with mobile colloidal materials. Soil solution studies generally show that plant response to metals is correlated with the free metal ion activity. Therefore, one aspect of metal bioavailability is related to which factor or factors contribute to the activity of the free metal ion in the soil solution. These interactions are summarized in Figure 1. Equilibrium models often are used to estimate free metal ion activities. The difficulty with the application of these models in the soil–root environment is associated with properly modeling all of the interactions identified above.

[12] G. M. Pierzynski and A. P. Schwab, in Proceedings of the Conference on Hazardous Waste Research, ed. L. E. Erickson, Manhattan, KS, 1990, pp. 511–520.

[13] D. F. Aoki, 'The Uptake of Arsenic and Cadmium in Mine Tailings by Poplar Trees', MS Thesis, University of Iowa, Iowa City, IA, 1992.

[14] US Environmental Protection Agency, 'Test Methods for Evaluating Solid Waste: Physical/Chemical Methods', SW-846. USEPA, Office of Solid Waste and Emergency Response, Washington, DC, 1986.

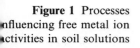

Figure 1 Processes influencing free metal ion activities in soil solutions

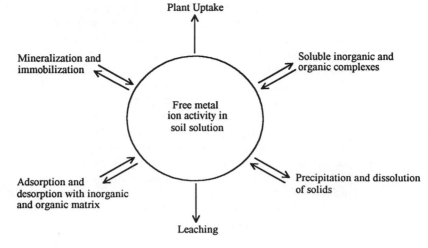

Metal fractionation or sequential extraction schemes sometimes are used to describe metal behavior in soils.[7] The schemes cannot be entirely specific for a given fraction within the soil, and an additional problem of re-adsorption of extracted metals to the soil constituents exists. Therefore, the value of the schemes in obtaining information on fundamental processes that influence metal behavior in soils is limited. However, the schemes can be useful in an empirical sense.

Remediation of a metal-contaminated site can include three possible changes in the chemical characteristics of the soil or mine spoil material. The total metal concentration can be reduced, as is the case with washing or leaching procedures. The TCLP concentration can be reduced without removing any of the metal, as is the case with solidification or vitrification processes. The metal bioavailability can also be reduced. *In situ* methods for reducing bioavailability include sorption, ion exchange, precipitation, and attenuation.[15] Increasing soil pH also has been evaluated for cationic metals.[7] Little information has been published with regard to the effectiveness of the soil treatments other than data on yield and metal concentrations in plant tissue.[16,17] In particular, detailed studies of the effects of soil amendments on free metal ion activities have not been reported. This is partly due to a lack of the necessary thermodynamic data.

3 Microbial Aspects of Metal-contaminated Soils and Mine Spoils

Plants may accumulate as much as 10 000 mg of Zn or 2500 mg of Pb per gram of shoot biomass.[18] Heavy metal tolerant plant species which concentrate and

[15] R. Sims, D. Sorensen, J. Sims, J. McLean, R. Mahmood, R. Dupont, J. Jurinak, and K. Wagner, 'Contaminated Surface Soils In-place Treatment Techniques', Noyes Publications, Park Ridge, NJ, 1986.

[16] M. S. Johnson, T. McNeilly, and P. D. Putwain, *Environ. Pollut.*, 1977, **12**, 261–277.

[17] W. E. Sopper, *Landscape Urban Planning*, 1989, **17**, 241–250.

[18] A. J. M. Baker, *J. Plant Nutr.*, 1981, **3**, 643–654.

detoxify metals in above ground plant parts are known as accumulator species. Detoxification mechanisms for these species may include binding of heavy metals to cell walls, pumping heavy metal ions into vacuoles, or complexing of heavy metals by organic acids. In contrast, excluder plants species may absorb heavy metals but restrict their transport into shoots. This type of heavy metal tolerance does not prevent uptake of heavy metals but restricts translocation, and detoxification of the metals takes place in the roots. Mechanisms proposed for excluder detoxification include immobilization of heavy metals on cell walls, exudation of chelate ligands, or formation of a redox or pH barrier at the plasma membrane.[19] Microbial immobilization of heavy metals in the root zone would also reduce availability to and uptake by plants.

In contaminated sites, heavy metal concentrations may be high enough to inhibit microbial activity. Soil micro-organisms may be critical to plant growth because they encourage development of a stable soil structure, release required nutrients in inorganic forms by mineralization, and produce growth-regulating substances. Also, soil micro-organisms may contribute to plant growth by immobilizing heavy metals in soil. The direct effects of Cd, Cu, Zn, and Pb on soil micro-organisms are generally understood.[20] Heavy metal contamination of soil decreases microbial activity, microbial numbers, and microbially mediated soil processes such as nitrification, denitrification, and decomposition of organic matter.[21–24] Higher numbers of resistance bacteria are found in heavy metal contaminated soil than in uncontaminated soil, and resistant communities isolated from long-term contaminated soils are more diverse than those found in recently contaminated soils.[25–28] However, at extremely high levels of contamination, fewer resistant bacteria have been isolated than from less polluted soils.[24]

Previous research has indicated that microbes can bind metals. Micro-organisms may accumulate metal ions by complexation with extracellular polymers,[29] or by ion exchange with polyanions of the bacterial cell wall.[30] Gram-positive bacteria have a greater ability to bind metals than Gram-negative bacteria due to cell wall structural differences,[31] although it has been suggested that the Gram-negative cell envelope acts to impede metal ion entry into the cell interior. Bacteria may be able to transform heavy metals by the production of

[19] G. T. Taylor, *J. Plant Nutr.*, 1987, **10**, 1213–1222.
[20] E. Baath, *Water Air Soil Pollut.*, 1989, **47**, 335–379.
[21] F. H. Chang and F. E. Broadbent, *Soil Sci.*, 1981, **132**, 416–421.
[22] A. Nordgren, E. Baath, and B. Soederstroem, *Soil Biol. Biochem.*, 1988, **20**, 949–954.
[23] J. M. Bollag and W. Barabasz, *J. Environ. Qual.*, 1984, **11**, 196–201.
[24] P. Doelman and L. Haanstra, *Soil Biol. Biochem.*, 1979, **11**, 487–491.
[25] B. H. Olsen and I. Thornton, *J. Soil Sci.*, 1982, **33**, 271–277.
[26] M. Kiroki, *Soil Sci. Plant Nutr.*, 1992, **38**, 141–147.
[27] K. G. Shetty, M. K. Banks, B. A. Hetrick, and A. P. Schwab, *Water Air Soil Pollut.*, 1993, accepted.
[28] T. J. Beveridge, in 'Metal Ions and Bacteria', ed. T. J. Beveridge and R. J. Doyle, John Wiley and Sons, Inc., New York, NY, 1989, 1–29.
[29] G. Bitton and V. Freihofer, *Microb. Ecol.*, 1978, **4**, 119–125.
[30] T. Rudd, R. M. Sterritt, and J. N. Lester, *Microb. Ecol.*, 1983, **9**, 261–272.
[31] T. J. Beveridge and S. F. Koval, *Appl. Environ. Microb.*, 1981, **42**, 315–335.

water-soluble organics which would increase metal solubility,[32] or release metals previously bound due to variations in metabolism or growth.[33]

The soil fungal population may similarly be affected by heavy metal contamination, with the diversity of micro- and macro-fungi decreasing in contaminated soils.[34] In the higher fungi, the production of sporophores is a sensitive measure of metal pollution.[20] One specific group of fungi, the mycorrhizal fungi, can directly contribute to plant tolerance of heavy metals. Mycorrhizal fungi are plant symbionts which proliferate inside and outside of host plant roots. The hyphal strands of the fungus exterior to the root absorb nutrients and translocate them into the plant. These fungi can bind metals to hyphae, restricting them from translocation to shoots.[35] To what extent mycorrhizal symbiosis affects heavy metal translocation patterns expressed by plants is not known.

The effect of vegetation on groundwater contamination by leachate from contaminated soils is uncertain.[36] The mobilization of biologically available metals may be slightly higher in vegetated soil[37] due to the release of complexing agents by the plant. The concentration of Zn in the leachate from contaminated mine tailings is higher in soils treated with 1.0 mM succinic acid than in the absence of organic acid (Table 2).[38] The adsorption of heavy metals to soil may also decrease in the presence of organic ligands found in the rhizosphere.[39] Plant roots may also influence water transport and metal movement by providing flow channels in the soil. Other research indicates that heavy metal leachate may be affected by the type of soil microflora associated with the plant (Table 3). Revegetation of heavy metal contaminated soil may increase heavy metal leaching, especially if soil microflora have not been fully restored.[40]

4 Vegetative Remediation at Mine Sites: Case Studies

Whitewood Creek

Revegetating mine sites and metal wastes offers several advantages that have been under-appreciated in the literature. Fast growing hybrid poplar trees have been used in a variety of climate zones in riparian area applications to stabilize soils, decrease wind-blown dust, and decrease vertical migration of pollutants. Most risk assessments at mine tailings sites indicate that the largest cancer risk for elements like As and the largest chronic health risk to humans from elements such as Cd are due to inhalation of wind-blown dust or ingestion of aeolian-deposited soil by children. Vegetation can decrease these exposure pathways

[32] A. J. Francis, S. Dobbs, and B. J. Nine, *Appl. Environ. Microb.*, 1980, **40**, 108–113.

[33] C. A. Flemming, F. G. Ferris, T. J. Beveridge, and G. W. Bailey, *Appl. Environ. Microb.*, 1990, **56**, 3191–4203.

[34] H. Yamamoto, K. Tatsuyana, and T. Uchiwa, *Soil Biol. Biochem.*, 1985, **17**, 785–790.

[35] R. Bradley, A. J. Burt, and D. J. Read, *New Phytol.*, 1981, **91**, 197–209.

[36] F. L. Domergue and J. C. Vedy, *Int. J. Environ. Anal. Chem.*, 1992, **46**, 13–23.

[37] J. M. Besser and C. F. Rabeni, *Environ. Toxicol. Chem.*, 1987, **6**, 879–890.

[38] M. K. Banks, C. Y. Waters, and A. P. Schwab, *J. Environ. Sci. Health*, 1993, accepted.

[39] P. Chairidchai and G. S. P. Ritchie, *Soil Sci. Soc. Am. J.*, 1990, **54**, 1242–1248.

[40] M. K. Banks, G. R. Fleming, A. P. Schwab, and B. A. Hetrick, *Chemosphere*, 1993, accepted.

Table 2 Average zinc concentration in the leachate of organic acid amended mine tailings[38]

| | Average Zinc Concentration in Leachate/μg l^{-1} Acid Concentration/μm | | | |
Type of Acid	0	50	250	1000
Formic	361	423	352	332
Succinic	362	308	492	506

Table 3 Average concentration of zinc leached from heavy metal contaminated soil by varying plant and microbial treatment[40]

Treatment	*Zinc*/mg l^{-1}
With plants	
Unamended	371
Microbes	228
Mycorrhizae	360
Microbes and Mycorrhizae	271
No plants	
Unamended	263
Microbes and Mycorrhizae	189

dramatically. Revegetation can be considered as a remediation method or used in tandem with other techniques for stabilizing soils and closing sites at low cost.

If contamination is in the upper 2–3 m of soil, deep-rooted poplar trees can significantly decrease the downward migration of leachate via evapotranspiration.[41] The trees start from 2 m 'whips', cuttings that have preformed root initials. When planted at a depth of 2 m, they form a dense root mass that will take up large quantities of moisture, increase soil suction, and decrease downward migration of pollutants. In the dormant season, some leakage of water can occur through the system but, precipitation is not great during this period. The trees grow 2 m in the first growing season and reach a height of 6–8 m after three years when planted at a density of 10 000 trees per hectare. Carbon fixation is approximately 2.5 kg m^{-2} yr^{-1}. Various management schemes can be adopted, and the trees can remain with very little attention for twenty years or more after the second season.

Advantages and disadvantages of vegetative remediation were discussed previously. Additional concerns specifically for trees include leaf litter and whether associated toxic residues might be blown off site. This concern may be tested in the laboratory or field to determine whether uptake and translocation of the metals into the leaves of trees or grasses exceed standards. In general, Cd and As (arsenate) are the most problematic because of their chemical similarity to nutrients (Ca, Zn, and P). Pb, Cr, Hg, and other metals are of lesser concern because of smaller rates of uptake. Following is a case study that illustrates an investigation of this potential problem at a Superfund site.

[41] L. Licht, 'Poplar Tree Buffer Strips Grown in Riparian Zones for Biomass Production and Non-point Source Pollution Control', PhD Dissertation, Civil and Environmental Engineering, The University of Iowa, Iowa City, IA, August, 1990.

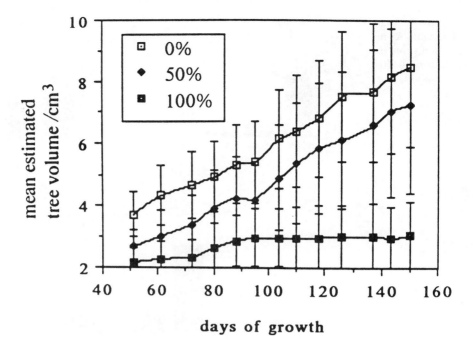

Figure 2 Cumulative growth curves for poplar trees in three fertilized laboratory treatments of 0, 50, or 100% mine tailings[13]

An eighteen mile stretch of Whitewood Creek is a US Superfund site because of contamination of surface water and groundwater with As and Cd (arsenopyrite is the major mineral in the tailings) from 130 years of gold mining activity. It is located in the Black Hills of extreme western South Dakota below the town of Whitewood. Chemical characterization indicated that the tailings contained an average of 1250 mg kg^{-1} total As and 9.4 mg kg^{-1} total Cd with pH ranging from 3.9 to 5.4. Plant-available P and K levels were quite low. An experimental plot was planted with 3100 hybrid poplar trees to a depth of 1.6 m in April of 1991. A commercial NPK fertilizer was used at recommended rates to ensure vigorous early growth of the cuttings. Roots formed along the entire length of the cutting in the soil, so a dense root mass was established that takes up infiltration and intercepts interflow moving towards the creek.

Genetically identical cuttings also were established in a plant incubator in the laboratory.[13] Figure 2 shows that the cuttings established in 100% mine tailings, the worst case from the site, grew more slowly than the other trees under optimal conditions in the laboratory. All trees were fed Hoagland R growth medium containing major nutrients. Other treatments were grown in a mixture of mine tailings and peat : vermiculite (50 : 50 by mass mixture). The treatment with 0% mine tailings was composed of a peat : vermiculite mixture, ideal for plant growth.

At the end of the first growing season, the trees had grown to 12 m at the field site. Leaves, stems, and roots were collected from the field as well as the laboratory trees to compare As and Cd uptake and translocation. Poplar leaves in the field did not accumulate significant amounts of As or Cd (Figures 3 and 4). These concentrations are below most levels established for field application of municipal sewage sludge or compost. Furthermore, they are well below the

Figure 3 Total acid-digestable As in leaves, stems, and roots in fertilized laboratory (0, 50, or 100% mine tailings) and field poplars[13]

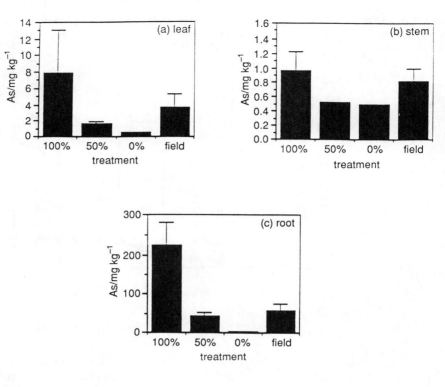

Figure 4 Total acid-digestable Cd in leaves, stems, and roots in fertilized laboratory (0, 50, or 100% mine tailings) and field poplars[13]

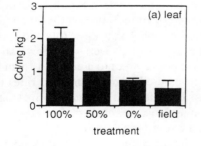

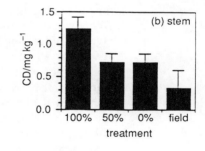

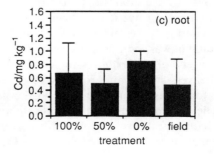

reference concentrations accumulated by leaves in the laboratory treatments of 100% mine tailings, indicating that the laboratory study overestimated the amount that would be accumulated in the field, possibly because of ideal growth conditions in the laboratory. It is interesting to note that the commercial peat : vermiculite mixture allowed a greater uptake of Cd by leaves than did the field situation (Figure 4a). Small amounts of Cd are always present in most commercial nursery mulches and soil amendments.

Concentrations of Cd and As in native vegetation at the site were generally of the same order of magnitude as those in poplar trees. But the leaves of lambsquarter were particularly high in As ($14\,mg\,kg^{-1}$), and the leaves of the native cottonwoods (a cousin of the hybrid poplar trees) have a somewhat higher concentration ($1.6\,mg\,kg^{-1}$) than leaves of the poplars planted for vegetative remediation. Results indicate that the poplars are not a serious concern in terms of their bioconcentration potential as compared with native vegetation at the site.

Laboratory and field investigations have shown that hybrid poplar trees can be established in mine tailings at a Superfund site without objectionable uptake of As and Cd into leaves. The laboratory study showed that estimates can be made easily and quickly regarding uptake and toxicity of metals. Field results demonstrated that the technology can be used at shallow contaminated sites for soil stabilization or in conjunction with other methods for closing a site.

Tri-State Mining Region

A number of studies have been made relating directly to vegetative remediation or to factors involved in establishing vegetation on contaminated soils or mine spoils in the Tri-State Mining Region. Several studies have dealt with a mine waste material known as chat, and one study examined a contaminated alluvial soil. Chat is a rock waste material generated from the initial processing of the metalliferous ore and consists primarily of rock fragments ranging in size from approximately $4\,mm$ to clay sized ($<2\,\mu m$) and having Zn, Pb, and Cd concentrations as high as 43 750, 4500, and $160\,mg\,kg^{-1}$, respectively.[12] Chat piles are scattered throughout the area. The finer sized particles are selectively eroded away from the piles by wind and water and contribute to metal-enriched sediments and wind-blown dusts.

The Galena Superfund site consists of a large area that is nearly void of vegetation and contains numerous piles of chat and other waste materials. The remediation plan calls for using the piles of material to fill in mine shafts and other voids, recontouring to control run-off, and establishing vegetation to further control erosion and run-off.

An unbalanced factorial arrangement of organic waste amendments (composted yard waste, composted cattle manure, spent mushroom compost, and turkey litter); organic waste application rate (0, 22.4, 44.8, and $89.6\,Mg\,ha^{-1}$); and inorganic fertilizer rate (zero, a rate recommended for native grass establishment, and a rate recommended for establishment for a grass–legume mixture) was used to evaluate vegetative responses in a chat material seeded with a mixture of native

Table 4 The effect of organic waste source and rate and of fertilizer rate on plant density, species richness, and total cover after amendment of a zinc–lead chat tailing[42]

Main effect	Plant density/ plants m^{-2}	Species richness (number of species)	Total cover/%
Organic waste source[a]			
TL	15.6a[b]	6a	35a
CM	86.3b	26b	44b
MC	57.9c	16c	36ab
YW	61.5c	21bc	41ab
C	33.0d	9a	10c
Organic waste rate (mg ha^{-1})			
0	32.9a	9a	10a
22.4	53.7b	21bc	32b
44.8	54.4b	18b	40c
84.6	57.9b	25c	45c
Fertilizer rate[c]			
none	57.5a	22a	40a
NG	55.4a	20a	35a
GL	47.9a	20a	35a

[a]TL = turkey litter, CM = composted cattle manure, MC = spent mushroom compost, YW = composted yard waste, C = control
[b]Means within the same column and main effect followed by the same letter are not significantly different at the 0.05 level
[c]NG = rate recommended for establishment of native grasses, GL = rate recommended for establishment of a grass–legume mixture

and tame grasses and leguminous forbs.[42] Table 4 shows the effects of organic waste source, organic waste rate, and fertilizer rate on total plant density, species richness, and total cover after the initial growing season. All three response variables were increased significantly by the organic waste sources as compared with the control, with composted cattle manure generally providing the greatest increase and turkey litter giving the least increase. The poor performance of turkey litter as compared with the other organic waste sources was due to acidification caused by nitrification of ammoniacal nitrogen forms in the material. Significant increases were also evident with increasing rates of organic waste. The addition of fertilizer had little beneficial effect, however. The combined results suggested that merely supplying the primary plant nutrients (N, P, and K) is not sufficient for acceptable establishment of vegetation in this material. Although the organic waste materials increased plant-available N, P, and K as well, they also increased organic C levels, cation exchange capacities, and the concentrations of other secondary and micronutrients (data not shown). Any potential benefits with regard to alleviating Zn phytotoxicity are unknown. This work has been applied directly to the remediation efforts at the Galena Superfund site.

[42] M. R. Norland, Proceedings of the Association of Abandoned Mine Land Programs, Dept. of Natural Resources, Div. Environ. Qual., Jefferson City, MO, 1991, pp. 251–264.

Table 5 The influence of organic and inorganic fertilizers and mycorrhizal fungi on biomass production and in uptake by *Andropogon geradii* and *Festuca arundinacea* grown in chat[43]

| | Fertilizer amendment | | | | | |
| | none | NH_4 | manure | KH_2PO_4 | NH_4 and manure | Manure and KH_2PO_4 |
Mycorrhizal treatment						
Biomass/g						
A. geradii[a]						
mycorrhizae	0.07bc	0.03c	0.08b	0.08b	0.41a	0.51a
no mycorrhizae	0.07bc	0.05bc	0.05bc	0.07bc	0.06bc	0.05bc
F. arundinacea						
mycorrhizae	0.07b	0.04b	1.04a	0.07b	1.43a	1.47a
no mycorrhizae	0.08b	0.07b	0.03b	0.07b	0.03b	0.06b
Zn uptake/mg plant^{-1}						
A. geradii[b]						
mycorrhizae	66c	nd	168bc	127bc	330ab	520a
no mycorrhizae	88c	78.8c	nd	75c	nd	95c
F. arundinacea						
mycorrhizae	159c	120c	905b	167c	1827a	984b
no mycorrhizae	241c	119c	84c	186c	294c	133c

[a]Means for each plant species followed by the same letter are not significantly different ($P = 0.05$)
[b]nd = not determined because of insufficient root biomass

The role of mycorrhizal fungi in establishing vegetation in the chat material also has been studied.[43] Table 5 shows the effect of various amendments and mycorrhizal fungi on biomass production and Zn uptake by big bluestem [*Andropogon geradii* Vit.] and tall fescue [*Festuca arundinacea* Schreb.]. Big bluestem is an obligate mycotroph that requires mycorrhizae to grow in soils with low fertility, whereas tall fescue is a facultative mycotroph that grows well in low fertility environments in the absence of mycorrhizae. In this situation, additional biomass production occurred only when mycorrhizae were present with adequate nutrients, illustrating the importance of fungi in alleviating Zn toxicity to the plants. The exact mechanism for this is not known. It may be related to the binding of the metals in the rhizosphere by the fungi or a change in the metal binding capacity of the cell walls, both of which could act to increase plant resistance to Zn.

Figure 5 shows the effects of various soil amendments on changing Zn bioavailability and the resulting changes in soybean [*Glycine max* (L.) Merr.] tissue composition and yields in a metal-contaminated alluvial soil. This soil was collected approximately 125 m from the Spring River in the Tri-State mining region and was in a field under crop production. Soybeans growing on site were severely chlorotic, and Zn phytotoxicity was the suspected cause because of high Zn concentrations (1090 mg kg^{-1}) in soybean tissue samples collected there. No mining activity had occurred adjacent to the field and the source of the Zn was

43 B. A. D. Hetrick, G. W. T. Wilson, and D. A. H. Figge, *Environ. Pollut.*, 1993, accepted.

Figure 5 Relationship between soybean tissue Zn concentrations and relative yield in a metal contaminated alluvial soil. The variation in tissue Zn concentrations was a result of changes in soil bioavailable Zn levels induced by various soil amendments without changing the total Zn concentration[7]

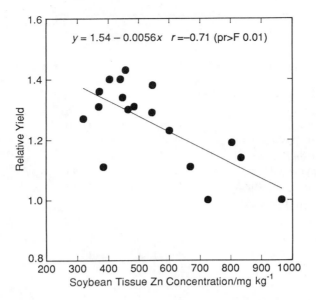

$y = 1.54 - 0.0056x$ $r = -0.71$ (pr>F 0.01)

metal-contaminated sediments deposited during periodic flooding events. The amendments were lime, P, cattle manure, sewage sludge, poultry litter, or various combinations of lime and cattle manure. The amendments produced KNO_3^- extractable Zn concentrations from 3.7 to 63.3 mg kg^{-1} with a corresponding range of soybean tissue Zn concentrations of 318 to 1153 mg kg^{-1}. Soybean yields were influenced by the changes in tissue Zn concentrations with a range of 1.0 to 1.4 (Figure 5). The manipulation of bioavailable Zn levels was done without changing the total Zn concentration of the soil. Although the overall thrust of this project was not vegetative remediation, it is one of the few studies that provides data on soil chemical changes induced by soil amendments designed to reduce metal bioavailability.

Studies have shown that various amendments and mycorrhizal fungi aid in enhancing plant growth in contaminated soils and mine spoils from the Tri-State Mining Region. As a result, a vegetative remediation strategy is being used at the Galena Superfund site as part of the overall clean-up effort. Additional information has been obtained regarding the importance of mycorrhizae for plant growth under Zn toxic conditions and on the usefulness of soil chemical fractionation schemes in assessing soil chemical changes induced by amendments.

5 Modeling of the Fate of Heavy Metals in Vegetated Soils

The root–soil water transfer process is a major part of the sub-surface hydrologic system. The development of quantitative models that describe water movement in the root–soil environment has been reviewed.[10] The Leaching Estimation and

Chemistry Model, LEACHM, has been used to simulate the movement of water and solutes through both layered and non-layered soil profiles.[44] A coupled root–soil water flow model that includes the vertical movement of water through the root system has been developed.[45,46] Soil water movement in the vertical and horizontal directions of a non-homogeneous variably saturated soil can be simulated with this model.

Some of the processes that occur in the soil–root environment are limited by the rate of diffusion or reaction and kinetic models should be used rather than the equilibrium models described earlier. Diffusion within solids is slow; it is often an important consideration when modeling the leaching of metals in soil.

Two important considerations in modeling the fate of metals in the root–soil environment are the uptake into the plant and the impact of root exudates on pH and leaching. Because micro-organisms degrade root exudates, any modeling of the impact of the organic ligands on metal leaching should include a root exudate and a microbial population balance.[47]

When the behavior of the solute is modeled with an equilibrium model, two distinct cases can be considered. Below the solubility limit, the metal will not precipitate, and a precipitated solid phase will not be present. On the other hand, when a precipitated solid phase is present, the solute concentration will be at the solubility limit and will remain at that value until all of the solid phase is dissolved. In this case, a model for the solid phase is needed to simulate the dissolution process and follow the transient behavior of the mass of precipitated metal. In the model that follows, the first case is considered.

As discussed previously, a variety of factors govern the fate of heavy metals in a vegetated soil; however, providing detailed mathematical expressions describing all of these processes would produce a nearly intractable problem. Thus, a somewhat simplified approach will be employed for developing a method to predict the fate of heavy metals in a rooted soil. Figure 6 depicts the conceptual approach used in development of the fate and transport model. It is assumed that the primary mechanism for metal transport through a soil is water movement, with losses or additions of metals to the soil–water occurring from four sources: (1) uptake into the vegetation's root system by plant transpiration; (2) adsorption onto the vegetation's root system; (3) bonding to exchange sites on inorganic soil constituents; and (4) adsorption to insoluble soil organic matter.

A model that has been shown to provide an accurate depiction of the movement of water in the presence of a transpiring crop's root system can be described as:[45,46]

$$\frac{\partial}{\partial z}\left[K_s \frac{\partial(\psi_s + z)}{\partial z} \right] - q = \left[\beta S_s + S_y \frac{dS_e}{d\psi_s} \right] \frac{\partial \psi_s}{\partial t} \tag{1}$$

[44] R. J. Wagenet and J. L. Hutson, 'Leaching Estimation and Chemistry Model, Version 2.0, A Process Based Model of Water and Solute Movement, Transformations, Plant Uptake, and Chemical Reactions in the Unsaturated Zone', *Continuum*, Vol. 2, Water Resources Institute, Cornell University, Ithaca, NY, 1989.

[45] M. A. Marino and J. C. Tracy, *J. Irrig. Drain. Eng.*, ASCE, 1988, **114**, 588–604.

[46] J. C. Tracy and M. A. Marino, *J. Irrig. Drain. Eng.*, ASCE, 1989a, **115**, 608–625.

[47] L. C. Davis, L. E. Erickson, E. Lee, J. F. Shimp, and J. C. Tracy, *Environ. Prog.*, 1993, **12**, 67.

G. M. Pierzynski et al.

Figure 6 Schematic representation of the modeling approach

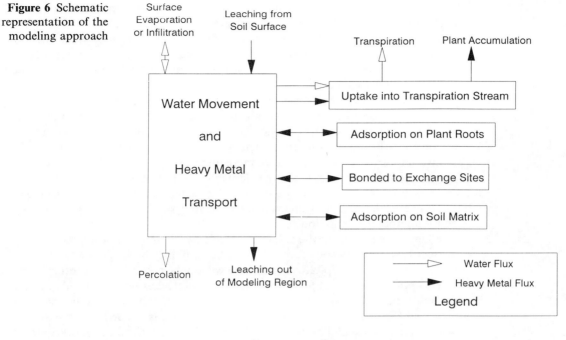

$$\frac{\partial}{\partial z}\left[K_r \frac{\partial(\psi_r + z)}{\partial z} \right] + q = R_d \frac{\partial WC_r}{\partial t} + WC_r \frac{\partial R_d}{\partial t} \tag{2}$$

in which z is the vertical direction in the soil; K_s is the hydraulic conductivity of the soil in the vertical direction; K_r is the hydraulic conductivity of the root in the vertical direction; ψ_s is the soil–water pressure head; ψ_r is the root–water pressure head; S_y is the specific yield of the soil; S_s is the specific storage of the soil; $\beta = 0$ if $\psi_s < 0$ and $\beta = 1$ elsewhere; WC_r is the root–water content, a function of the root–water pressure head; R_d is the root density in the soil; t equals time; $S_e = \theta/n$ which is the effective saturation of the soil, where θ is the soil–water content and n is the soil porosity; and q equals the rate at which soil–water is extracted by the plant's root system per unit volume of soil, defined as:

$$q = S_e R_d \Gamma (\psi_s - \psi_r) \tag{3}$$

where Γ is a lumped parameter representing the permeability of a plant's root system.

Equations (1) through (3) represent a coupled set of partial differential equations that can be solved numerically, given that the root parameters, the soil characteristics, the initial conditions, and boundary conditions are known.

The solutions of equations (1) through (3) describe the distributions of the water flux throughout the soil profile. Thus, Darcy's law is employed to calculate the water flux, V, distribution based on the soil–water pressure heads, such that:

$$V = - K_s \frac{\partial(\psi_s + z)}{\partial z} \tag{4}$$

The transport of heavy metals through the soil profile can then be described using the water flux distribution and the advection–dispersion equation, as:

$$\frac{\partial}{\partial z}\left[\theta D \frac{\partial C}{\partial z} - VC\right] - q_m + S = \frac{\partial}{\partial t}[C(\theta + K)] \tag{5}$$

in which C is the concentration of heavy metals in the soil–water; D is the macrodispersion coefficient for heavy metals in the soil; q_m is the uptake of heavy metals by roots into the plant transpiration stream; S is the sink/source of contaminants across modeling boundaries; and K equals a lumped parameter accounting for the adsorption onto root and soil surfaces and ion exchange.

Several heavy metals are necessary as plant nutrients (*e.g.* Cu and Zn)[48] and a significant fraction of these metals can be taken up by the roots during plant transpiration. However, other metals are toxic to some plant species and a larger fraction of these metals are excluded during the root uptake process. Thus, development of the uptake term, q_m, in equation (5) will have to be based on the specific plant type and metal being studied, and very few generalizations can currently be made about this process. Nonetheless, a general model that may be useful for developing a mathematical simulation model of this process is similar to a proposed model[49] for the uptake of organic chemicals into a plant's transpiration stream, simulated as a linear function of root water uptake, so that:

$$q_m = f_u q \tag{6}$$

where f_u equals the ratio of the concentration of the chemical in the root water to the concentration in the soil water. Use of this expression for metals in conjunction with equation (3) would allow the uptake to be calculated as:

$$q_m = f_u R_d S_e \Gamma(\psi_s - \psi_r) \tag{7}$$

where f_u would have to be calibrated as a site-specific parameter.

The lumped parameter, K, in equation (5) actually accounts for three processes: (1) adsorption to the root mass; (2) adsorption to the soil matrix; and (3) ion exchange. Very little information is available to quantify these relationships based on the general characteristics of a site. However, it is felt that a reasonable approximation would be to assume that the heavy metals partition into each phase (water, root, soil, and biomass) in a linear fashion and that the time frame of the simulations is of long enough duration to assume equilibrium conditions. In this fashion, the lumped adsorption parameter, K, can be described as:

$$K = \rho k + R_d k_r + k_e E \tag{8}$$

in which ρ is the soil density; k equals the linear partition coefficient between the soil–water and soil matrix; k_r is the linear partition coefficient between the soil–water and the root mass; k_e is the linear partition coefficient between the soil–water and the ion exchange sites on the solid phase; and E is the cation exchange capacity of the soil.

The concentration C includes all forms of the metal dissolved and suspended in the soil–water, and can be written in terms of its components: C_1, the

[48] A. J. Friedland, in 'Heavy Metal Tolerance in Plants: Evolutionary Aspects', ed. A. J. Shaw, CRC Press, Boca Raton, FL, 1989, pp. 7–20.

[49] G. G. Briggs, R. H. Bromilow, and A. A. Evans, *Pestic. Sci.*, 1982, **13**, 495–504.

concentration of the charged species; C_2, the concentration of metal complexed with an inorganic species, A_2; and C_3, the concentration of metal complexed with organic species, A_3.

The total concentration, C, can then be expressed in terms of C_1 as:

$$C = C_1 + K_2 A_2 C_1 + K_3 A_3 C_1 \tag{9}$$

where K_2 and K_3 are the partition coefficients associated with the metal complexed with the inorganic and organic species, respectively.

If the ion exchange process involves only the charged species C_1, then equation (5) may be written in the form:

$$\frac{\partial}{\partial z}\left[\theta D \frac{\partial[C_1(1 + K_2 A_2 + K_3 A_3)]}{\partial z} - VC_1(1 + K_2 A_2 + K_3 A_3)\right] - f_u R_d S_e \Gamma(\psi_s - \psi_r) + S$$

$$= \frac{\partial}{\partial t}[C_1(1 + K_2 A_2 + K_3 A_3)(\theta + \rho k + R_d k_r) + C_1 k_e' E] \tag{10}$$

It is likely that the root exudate concentration, rhizosphere biomass density, and root density probably will be variable over time. Thus, some mechanism of predicting these densities as they vary in relation to the time of year, climatic conditions, and general site conditions must be employed. However, such models will not be included here due to the limited scope of this paper.

The estimation of the soil matrix partition coefficient, k, in equation (8) also must be done on a site-specific basis and will depend on factors such as the soil's organic content and pH. With the introduction of vegetation at a site, both the organic content and the pH of the soil could change significantly and, thus, the partition coefficient must be described as a function of both, such that:

$$k = f(\text{pH}, \%Oc) \tag{11}$$

where $\%Oc$ is the percent organic matter in the soil matrix. In most cases it should be possible to develop this relationship on a site specific basis. However, a method also would have to be developed to predict the pH and organic content of a soil once vegetation has been introduced, which could prove difficult.

The model governing the fate and transport of a heavy metal can be solved numerically. The model solutions could then proceed by solving equations (1) through (3) to determine the water pressure distribution in the soil profile, then solving equation (4) to calculate the soil–water flux distribution. Finally, equation (10) could be solved using the soil–water flux and root water extraction calculations to determine the heavy metal concentration of the soil–water in the soil profile for each time increment during the simulation period. Equation (10) also could be solved simultaneously with the balances for root exudates and microbial biomass, if transient changes in root exudate concentration, A_3, are to be included in the model.

One of the challenges of using mathematical models to simulate the fate of heavy metals in soils is to collect sufficient equilibrium and soil characterization data. Although some data for root exudates has been collected,[50] more research

---

[50] A. P. Schwab and M. K. Banks, Proceedings of the 86th Annual Meeting and Exhibition of the Air and Waste Management Association, Denver, CO, June 14–18, 1993, Paper 93-WA-8906.

is needed before the model could be utilized to design a vegetative remediation scheme. However, the proposed model can be used to demonstrate the qualitative effects that the introduction of vegetation would have on soils contaminated with heavy metals.

Equations (1) and (2) simply describe the water flow regime in a variably saturated soil, with equation (4) providing a calculation of the water flux and equation (3) representing the amount of water extracted by a plant's root system. Any transpiration by vegetation introduced into a barren soil would result in the extraction of soil–water by plant roots. This would have two significant effects on the transport of heavy metals through a soil. First, the water sink provided by the plant roots generally would be strongest in areas with the largest root densities, typically near the soil surface. This would decrease the downward rate of water flow through the soil, thereby decreasing the mass of heavy metals leached below the root zone. Second, the water sink provided by the root system also would reduce the overall soil–water potential, ψ_s, which in turn would lower the soil–water content. Then, because the hydraulic conductivity of a soil decreases with decreasing soil–water content, the presence of the plant's root system also would decrease soil permeability, further restricting the movement of water and metals. Thus, equations (1) through (4) tell us that the introduction of a vegetative system to a heavy metal-contaminated soil would result in a type of hydraulic containment system.

Equation (5) describes the fate and transport of heavy metals through the soil profile. The advective and dispersive terms, V and D, are related to the movement of water. Equations (6) and (7) describe the uptake of heavy metals by a plant's root system. Because of their formulation, some metals probably would be taken up during the root–water uptake process described in equation (3), thereby reducing the mass of heavy metals in the soil profile. Thus, the introduction of vegetation would result in a reduction in the mass of heavy metals in a soil. However, the heavy metals that are taken up would accumulate in the plant biomass, typically in plant leaves. If the vegetative system were left unmanaged, plant dormancy at the end of a growing period would result in the decomposition of parts of the plant and the introduction of the heavy metals back into the soil at the soil surface where the leaves and other matter would fall, thus producing a source of metals, defined by the term S in equation (5). For the uptake to provide a true sink of heavy metals in the soil system, the vegetation would have to be managed in some way, such as harvesting, to prevent the reintroduction of the metals into the soil profile.

Equations (8) through (11) describe the adsorption of heavy metals to the organic and inorganic matter in the soil matrix. In general, heavy metals tend to adsorb readily to organic matter. The introduction of vegetation at a site would produce an increase in organic matter from the soil matrix to the plant's root system and the microbial mass associated with the plant's rhizosphere. Equation (8) thus indicates that the introduction of vegetation in a soil would decrease the mobility of the heavy metals, thereby providing a better containment system. However, most plant roots produce root exudates, so that a healthy environment is maintained in the rhizosphere for microbial and root growth. This property tends to alter the pH of a soil, so that root growth can be maintained at optimal levels, with pH values in the range of 6 to 8 favoring most plants.[51] If the natural pH of the contaminated soil is above these levels, the introduction of vegetation could result in a substantial lowering of the pH. This could reduce the partition coefficients in equation (9) and result in the metal becoming more mobile in the soil. In situations where the soil pH is below 8, soil amendments could be utilized to maintain a relatively constant pH, thus preventing an increase in the heavy metal mobility once the vegetative containment system is fully developed.

The modeling results that could be expected for the development of a vegetative remediation system would be extremely site dependent. However, the overall analysis of the model presented above suggests that, for many heavy metal contaminated soils, vegetation should provide a positive influence for enhancing the on-site containment of the metals and the possible removal of a portion of the metals through the harvesting of the vegetation.

6 Conclusions

Numerous remediation options exist for metal-contaminated sites. These range from complete excavation of contaminated material accompanied by some treatment to *in situ* encapsulation and to vegetative remediation. Vegetative remediation is aesthetically pleasing and it offers several advantages, including the minimization of erosion, low cost as compared with other remediation options, and the potential to reduce net percolation through contaminated sites. Amendments to contaminated soil or mine spoil materials may reduce metal uptake by plants by reducing metal bioavailability. Theoretically, such amendments likely reduce free metal ion activities in the soil solution although it is difficult to estimate or measure the actual treatment effects. On a more practical basis, chemical fractionation schemes are useful for quantifying treatment efficacy. Mycorrhizal fungi play an important role in establishing vegetation by allowing plants to utilize plant nutrients more efficiently and by decreasing plant sensitivity to phytotoxic metal concentrations. Thus, more contaminated areas may be suitable for vegetative remediation than was previously believed.

The use of trees holds particular promise for vegetative remediation. In addition to providing erosion protection, they have the potential to transpire considerable amounts of water compared to non-woody plant species. This may help in reducing the downward migration of contaminants. Trees can also produce biomass for chemical and/or energy use. Initial results suggest that food-chain transfer of contaminants due to uptake into leaves and stems is not a concern.

A model has been presented that can estimate the effects of vegetation on the fate of metals in contaminated soils and mine spoils provided the appropriate parameters can be obtained. The model takes into account root and soil characteristics, water balance, and the influence of vegetation on certain soil chemical properties with time. Use of the model would allow a more thorough appreciation and understanding of vegetative remediation.

---

[51] G. B. Tucker, W. A. Berg, and D. H. Gentz, in 'Reclaiming Mine Soils and Overburden in the Western United States: Analytical Parameters and Procedures', ed. R. D. Williams and G. E. Schuman, Soil Conservation Society of America, Ankeny, IA, 50021, pp. 3–26.

Acknowledgements

The work presented here was partially supported by the US EPA under assistance agreements R-815709 and R-919653 for the Great Plains–Rocky Mountain Hazardous Substance Research Center for regions 7 and 8. It has not been submitted to the EPA for peer review and therefore may not reflect the views of the agency and no official endorsement should be inferred. The US Department of Energy, Office of Environmental Restoration and Waste Management, Office of Technology Development, the Kansas Agricultural Experiment Station, and the Center for Hazardous Substance Research also provided partial funding.

Green Coal Mining

D. J. BUCHANAN AND D. BRENKLEY

1 Introduction

Coal now faces increasing environmental constraints in all facets of the fuel cycle. Furthermore, the business of coal is compounded by alternatives which are gaining ascendancy in world energy scenarios. It is a formidable task to ensure that coal is mined and used in an economically efficient and ecologically sustainable manner, particularly in those countries and regions which cannot presently meet the costs of environmental protection.

This paper sets out to show that coal can be mined in an environmentally acceptable manner, responsive to the global demands of ecological sustainability and energy supply security, and equally importantly, with diminishing hazards to the workforce. After setting the scene, the paper touches on the following topics: (1) review of mining methods, opencast and underground; (2) the importance of controlling workplace environmental hazards; review of hazards and controls; (3) mining environmental impacts and controls; surface and underground mining; (4) near-, medium-, and long-term technology themes to mitigate environmental impact; quality control and automation, product refinement, and *in situ* conversion, respectively.

Oil and gas aside, coal is the most important extractive industry in global terms. Worldwide, about 4.5 billion tonnes coal, all types, are mined annually with a net value of approximately \$175 billion.[1] The world's coal industry is large and expanding, with an average growth rate of about $2\frac{1}{4}\%$ p.a. throughout the 1980s. There is a significant export trade in hard coal, totalling around 0.4 billion tonnes p.a. which is increasingly influenced by requirements for low ash and sulfur content. The value of UK mineral production in 1988 was £15.3 billion, or a little under 4% of gross domestic product (GDP). Coal dominates with some £4.3 billion in value.

For the next decade and beyond, ecological sustainability will be one of the principal measures against which economic activities, including the mining of coal and its subsequent utilization, will be assessed. This will pose considerable challenges to those involved in the extraction and consumption of coal and other fossil fuels.

---

[1] P. C. F. Crowson, 'An International Perspective on the UK Coal Industry', *Min. Technol.*, July/August 1992, 215–221.

World recoverable coal reserves are estimated to be of the order of 850–1000 billion tonnes. Global growth in coal consumption will probably be of the order of 1% p.a. over this decade, with strongest growth in developing regions.[2] Coal's contribution to the world primary energy supply pattern is projected to be 27–30% over the next three decades. Consumption will be predominantly by the electricity generation sector with growth rates of 6% p.a. expected in developing countries. Coal presently fuels 40% of world electricity production.

Environmental and climate protection legislation, together with response strategies, will increasingly impact on the coal fuel cycle. This is exemplified by the following legislation: (1) the UN Framework Convention on Climate Change aimed at returning emissions of CO_2 and other greenhouse gases to 1990 levels by the year 2000; (2) the EC Large Combustion Plants Directive (88/609/EEC) and its implementation in the UK via the 1990 Environmental Protection Act; (3) renegotiations of UN ECE Protocol on Transboundary Air Pollution regarding emissions of sulfur and nitrogen oxides.

Legislation on SO_2 emissions is now in place in economically developed countries. The adoption of the critical loads approach to emissions control is gaining ground in Europe. Adoption of more stringent SO_2 emission limits in non-OECD countries is also evident.

Added to the above, Governmental energy policy formulation is likely in the longer term to include an assessment of external social and environmental costs and embody the principles of Least Cost Planning.

2 Mining Methods and Impacts

This section introduces the principal coal mining methods employed and associated environmental concerns.

Surface Mining

Surface, strip mining, or opencast mining is practised where the coal seam is nearly horizontal, covered with a relatively thin overburden, and where the surface topography is of low relief. The thickness of the overburden that can be removed economically depends on: (1) the recoverable coal resource value; (2) the thickness and resistance of the overburden to removal; and (3) the extractive plant available. Coal is mined by blasting the overburden and using draglines, shovels, and dumptrucks in what is essentially an earth-moving operation. A number of stripping techniques are available, but it is important to replace the overburden in the original sequence to avoid stratigraphic water problems. Hydraulic stripping is also used for overburden removal, principally in Russia. However, the land required for hydraulic waste dumps, difficulties in subsequent reclamation, and high specific water consumption are major demerits of hydraulic mining.

Surface mining currently contributes 30% of aggregate global hard coal production of approximately 3.3 billion tonnes although there is wide variation

---

[2] OECD/IEA, 'World Energy Outlook to the Year 2010', OECD/IEA, Paris, 1993.

Figure 1 Room and pillar
mining method

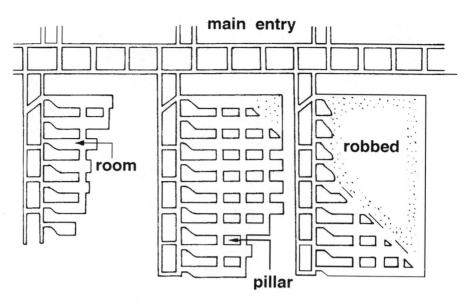

in national industry contribution. Surface mining methods have been adapted to a wide range of climatological conditions, ranging from equatorial to arctic zones.

Novel hybrid extraction methods have been proposed whereby tunnels are driven into a seam directly from an opencast highwall. This method provides an attractive option for opencast mines working at their maximum economic depth and bridges the transition to total underground commitment.

The advantages of opencast coal mining can be summarized as follows:

(1) Economic: high productivity, low production costs; reduced capital investment; reduced project execution time, 3–4 years maximum *versus* 10 years for an underground operation; reduced errors in economic planning; low labour intensity.
(2) Human: workplace less arduous; reduced health and safety risks.
(3) Technical: close control of progression of working; simple systems, high capacity machinery; operator control of workspace (absence of subsidence, *etc.*); derelict land restoration capability.

Underground Mining

Where the coal is deeply embedded, then a vertical shaft or a slope mine entry is constructed to the coal and lateral entries are driven into the coal bed. There are two generic extraction techniques; room and pillar (also called bord and pillar, pillar and stall, *etc.*) and longwall methods. Figures 1 and 2 illustrate the generic forms.[3]

Room and Pillar.   A number of room and pillar methods exist. Early schemes operated in a cyclic manner, *i.e.* drill–blast–muck. The trend, however, is towards

[3] T. Atkinson and M. J. Richards, 'UK Coal Mining in Prospect', *Min. Technol.*, August 1989, 249–254.

Figure 2 Longwall mining method

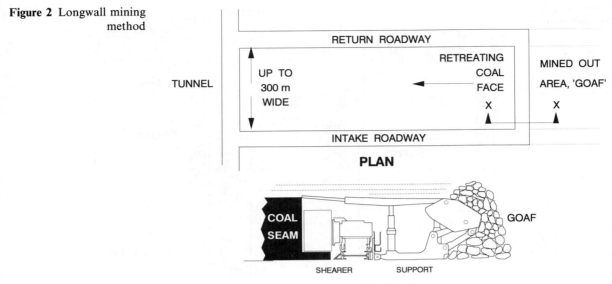

continuous schemes where a continuous mining machine stays in one tunnel or heading for some distance, often concurrently undertaking mechanized roofbolting. Mineral is transported by shuttle cars or mobile conveyors the trend again being towards continuous coal clearance systems. Where subsidence can be tolerated the pillars can be extracted on the retreat, permitting up to 90% extraction of the coal in place. At depth, required pillar size increases substantially.

Longwall Methods.   Longwall methods are used in medium-deep operations and are essentially of the advancing face or retreating face configuration. Most UK longwalls are now retreat; ten years ago most were advance. The UK coal face strategy has also focused on the adoption of high reliability, heavy duty face equipment, and shield supports. The success of the strategy of concentrating production from a minimum number of heavy duty retreat faces is critically dependent on the development and scheduling of replacement faces with well qualified geotechnical risk. This demands rapid roadway excavation rates, best achieved by continuous miners with concurrent excavation and support activities. Simple effective ground control with minimum convergence and requiring minimum maintenance and remediation is also a prerequisite.

Underground or Surface Mining?

The choice between surface and underground mining as the optimum mining method can no longer be based primarily on mining, geotechnical, and economic considerations.[4] An environmental (impact) assessment is now a prerequisite of many national legislatures, possibly supported by some form of public inquiry.

[4] M.S. Lindsay, *Aust. Coal J.*, 1985, **9**, 9–14.

Environmental assessment was given legal effect in the UK planning system through the Town and Country Planning (Assessment of Environmental Effects) Regulation of 1988 in response to the European Community Directive 85/337/EEC.[5] British Coal has participated in environmental impact assessments for several years, producing an assessment for every planning application. The earliest national legislative response was probably the National Environmental Policy Act passed in the United States in 1969. This, however, led to extensive confrontation and delays. The concept of Integrated Environmental Management with the structured parallel development of project and environmental management throughout all phases of a project is seen as a mechanism to avoid an adversarial approach and minimize litigation.[6] Risk assessment based approaches to the evaluation of surface *versus* underground mining and the environmental planning of mines are also emerging.

At the operational level environmental audits[7] are increasingly viewed as an effective tool in managing regulatory requirements and reducing potential liabilities. It is evident that increasing environmental regulation, liabilities, and associated engineering controls will dictate an increasing use of environmental professionals in the minerals industry.

Environmental Issues

Environmental impacts have been defined for the various stages of the coal fuel cycle.[8] The 'coal cycle' comprises five main activities: (1) exploration and extraction; (2) preparation; (3) handling and supply; (4) conversion (where applicable); (5) utilization, including waste disposal.

The principal environmental impacts and concerns specific to the exploration, extraction, and preparation phases are listed below:

Surface Mines—

> siting; large-scale land use; overburden removal and disposal; disturbance of hydrology and run-off; acid mine drainage; visual intrusion; noise, blast vibration, flyrock; fugitive dust; transportation/traffic; highwall stability; restoration of soil fertility; recreating ecosystem diversity; recreating landscape, amenity value; historic resource preservation.

Underground Mines—

> siting; spoil disposal: (a) magnitude of problem, (b) tip stability, and (c) tip combustion; lagoon requirements; subsidence; aquifer disturbance; minewater drainage/disposal; methane emissions; fugitive dust; visual intrusion; noise.

[5] A. Hambridge, N. Hill, and M. Caine, 'Environmental Assessment—The Need for Good Practice', *Mine Quarry*, April 1993, 16–17.

[6] J. S. Freer, 'Integrated Environmental Management in the Mining Industry', *J. S. Afr. Inst. Min. Metall.*, January 1993, 21–24.

[7] J. N. Philbrook, 'Environmental Audits: Determining the Need at Mining Facilities', *Min. Eng.*, February 1991, 207–209.

[8] UK Department of Energy, 'Coal Research, Development and Demonstration in the UK—The Way Forward', A Report from the Coal Task Force to the Department of Energy, June 1991.

Abandoned Mines—

methane migration; flooding; groundwater contamination; structural integrity; land rehabilitation.

Underground (Workplace) Environment—

ventilation—methane dilution, pollutant dispersal, temperature, and humidity control; pollutant sources and monitoring—$CO/NO_x/CO_2/O_2$ deficiency/diesel engine emissions/respirable dust/quartz/radon/occupational chemicals; spontaneous combustion/fire detection; dust explosibility; noise, vibration, illumination; heat stress management; mine drainage, in-rush protection; rockburst/microseismic activity, falls of ground; ergonomic hazard control.

3   Review of Workplace Environmental Hazards and Controls

The Need for Controls in the Workplace Environment

The earliest professional mining engineers recognized the association of industrial hazards and diseases with exposure to environments unfit for workers. Britain's historical lead in the intensive exploitation of its coal deposits led to early state intervention in the regulation and inspection of mines. It is, however, undeniable that the progression of legislation of health, safety, and environmental controls was reactive if not disaster-led. State regulation in UK coal mines can be traced back to 1842 but this was proceeded by much public pressure. The explosion at Felling Colliery, 1812, claiming 92 lives, led to the establishment of the Sunderland Society, resulting in the development of the Flame Safety Lamp, an important early milestone.

A number of Acts and Supplementary Acts can be linked as a response to particular disasters over time; for example the **Hartley Colliery Disaster**, 1862, with 204 entombed, resulting in a Supplementary Act requiring two means of egress, and **Aberfan**, with 109 killed, leading to the Mines and Quarries (Tips) Act of 1969 and Mines and Quarries (Tips) Regulations, 1971.

It was, however, not until 1974 with the arrival of the Health and Safety at Work Act that this reactive cycle of legislative formulation was effectively broken. This has progressed to the present with the Management of Health and Safety Regulations, 1993, which require an assessment of risks and actions to address them. In this latest legislation the importance of risk assessment is stressed. Proper consideration of hazards and adoption of appropriate methods to reduce their severity and frequency is fundamental to reducing injury and asset loss. The continuous diligence exercised by those involved in the UK mining industry towards securing a safe workplace is reflected in international accident statistics.[9]

[9] M. Widdas, 'British Coal's Safety Record—Fact or Fiction', Proceedings of the IME Symposium 'Safety, Hygiene and Health in Mining', Harrogate, UK, 18–20 November 1992, 14–27.

Year	Progressive massive fibrosis/%	All categories of pneumoconiosis/%
1954	1.0	13.5
1964	1.5	12.4
1969	1.3	12.0
1974	0.8	9.0
1978	0.4	5.2
1982	0.2	3.0
1986	0.0	0.9
1990	0.0	0.6

Dust

Evidence of the success of UK dust reduction measures is unequivocal.[10] Table 1 demonstrates the dramatic decline in the prevalence of pneumoconiosis over the last three decades. The reduction in progressive massive fibrosis, associated with the highest degree of disablement, is even more evident.

Research into dust formation mechanics and control methodologies has been taking place for over 30 years. Whilst a number of dust control methods have evolved over the years involving water sprays, air curtains, dust filters, and even foam suppression, without doubt one of the most effective of control technologies is the Extraction Drum.[11] Figure 3 shows a schematic view of the Extraction Drum. This is now seen as standard equipment on new coal-faces to reduce dust and frictional ignition. In recognition of this development British Coal gained its sixth Queen's Award for Technological Achievement in 1991.

Traditionally, the control of respiratory diseases associated with coal dust has been accorded the highest priority. There exist, however, other respirable hazards which are regarded with increasing concern. A number of national research programmes have been initiated in response. These hazards are discussed briefly in turn.

Quartz can cause the fibrogenic type lung disease, silicosis. A number of quartz exposure standards have been promulgated. In general, these standards must be referred to specific gravimetric dust monitoring instrument types. The analysis of quartz content in airborne dust samples is relatively complex. Several techniques are available for analysing quartz on gravimetrically collected samples; however, the perfection of a real time monitor remains elusive.

Natural ionizing radiation from solar and terrestial sources is part of our daily environment. The global average dose-equivalent rate from external sources[12] is about $0.8 \, \text{mSv a}^{-1}$. However, unlike toxic substances, there appears to be no

[10] D. A. Scarisbrick, 'Respiratory Disorders Associated with Coal Mining', Proceedings of the IME Symposium 'Safety, Hygiene and Health in Mining', Harrogate, UK, 18–20 November 1992, 117–129.

[11] K. Moses and D. J. Buchanan, 'The Extraction Drum—Less Dust and Fewer Ignitions at the Coalface', *Min. Technol.*, October 1991, 271–273.

[12] A. H. Leuschner and G. P. de Beer, 'Radiation Protection and Radon: Fundamentals, Techniques and Equipment', *J. S. Afr. Mine Ventilation Soc.*, March 1992, 43–51.

Figure 3 Schematic view of extraction drum

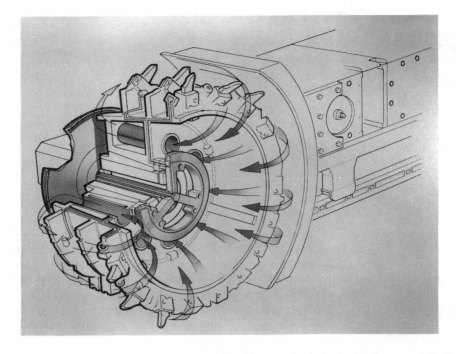

threshold below which there is no significant physiological effect or risk. It is currently accepted that for radiation doses associated with mining there is a linear relationship between exposure level and health risk and that consecutive doses are cumulative. A clear principle of striving to reduce mining associated doses to be 'as low as reasonably achievable' is thus widely accepted. For underground mineworkers three short-lived decay products of radon (radon daughters) pose the dominant radiation risk. The decay products of radon in air readily attach to aerosols and, on inhalation, can provide an important pathway for radiation dose. One of the principal problems confronting mining engineers is the accurate assessment of radiation exposure of workers. The measurement of radon gas in coal mine atmospheres is relatively straightforward. However, the measurement of radon radio-nuclides is complicated by their relatively short half-life, typically 30 minutes, and the complex radon gas/radon daughter concentration relationship. British Coal has undertaken extensive research to develop improved techniques for monitoring radon decay products and survey underground exposure. The use of carbon-loaded thermoluminescent dosemeters housed within existing gravimetric dust sampling instruments to determine time-averaged levels of decay products is considered an effective solution to satisfying current and anticipated legislation. Underground mitigation of radon is presently achieved by increasing ventilation; however, several other particle removal and strata sealing techniques have been proposed.

In recent years, increasing international concern has been expressed over the possible carcinogenicity of **diesel particulate matter** and various national regulations have been enacted covering the air dilution requirements and exposure limits for the gaseous exhaust pollutants of underground diesel engines.[13] The soot particles are almost entirely respirable in size, with 95% having mass median diameter less than 1.0 μm. A variety of organic materials are adsorbed on the carbon particles which subsequently can be deposited in the lung. In conjunction with high lung dust burdens, this may pose a small excess risk of lung cancer, and possibly bladder cancer. The causal mechanisms for cancer induction, dose–response relationships, and what constitutes a safe and feasible exposure limit are highly complex and contentious. Statutory exposure standards are under active consideration. The situation is complicated in that the measurement problem of discriminating diesel particulate matter from coal dust in mine air has not been solved, and control technologies are embryonic for flameproof diesels.

Noise

Noise-induced hearing loss is an insidious effect, associated with gradually accumulating damage. This has been addressed by a variety of legislation, culminating in the UK Noise at Work Regulations, 1989 and European Community Directive 86/188/EEC. This legislation places specific responsibilities on employers to appraise workers' operational environment, reduce noise whenever possible, and provide hearing protection.

Various national occupational exposure standards have been formulated. These have up to 10 dB differences in equivalent continuous sound level limits (Leq) for an eight hour shift.[14] In part, some of the difficulty relates to the wide susceptibility of individuals to hearing loss and the low correlation between temporary threshold shift values with permanent threshold shift.

British Coal's research into mining machinery noise, hearing protection, and noise exposure evaluation have been key elements in establishing UK noise policy and influencing how the European industries should comply with legislative requirements. Noise control research has been applied to machine hydraulics, electric motors, scraper conveyors, cutting systems, gearboxes, drilling pneumatics, and man-riding systems. Further research is required to control difficult sources such as drill rod noise and drum cutting noise.

Underground Fires and Ignitions

An underground fire constitutes a serious incident. Examination of the underlying causes of plant involved in underground fires suggests that mechanical failure of conveyor systems consistently accounts for the bulk of fires. Spontaneous combustion and ignitions are other significant causes.

Given the gravity of fire incidents, every reasonable measure is taken to reduce operational fire hazards. For example, fire-resistant hydraulic fluids are used wherever possible within the coal mining industry. In less demanding temperature and lubrication regimes emulsions are used. For extremes of heat, non-aqueous fluids are dominated by phosphate esters. These can be up to eight times the cost

[13] R. W. Waytulonis and D. J. Johnson. 'International Regulation of Diesel Engine Use Underground: A Country-by-Country Synopsis', US Bureau of Mines, Information Circular 9121, 1987.

[14] A. Hughes, 'Noise and the Law', *Colliery Guardian*, November 1989, 356–358.

of mineral oils. In selecting a material for its fire resistance it is also necessary to establish that there are not unacceptable toxicological risks to the workforce.

Mechanical and electrical engineering controls over sources of incendive energy are ubiquitously applied underground. Electrical systems in UK coal mines are designed to comply with flameproof and intrinsically safe standards (EN 50 014 through EN 50 020). Mechanical controls include the selection of fire-resistant materials, constraining external surface temperatures, employing local auxiliary ventilation, use of high pressure water sprays, avoidance of dry cutting, using sharp bits and reducing cutting tool speed, and exercising care in siliceous rocks and those containing inclusion of iron pyrite.

In the event of a fire where extensive spontaneous combustion is present, inert gases are used to assist in firefighting.

Suppression of dust explosions in coal mines is accomplished by the use of periodically located passive or triggered barriers using water or non-flammable dust as the extinguishing medium. Coal-face machine ignition suppression devices have also been developed. The principal explosion prevention measure, after ventilation, is the regular stonedusting of mine roadway floors to ensure a high inert material content.

Ventilation and Methane Control

The need to control 'firedamp' has historically been accorded the highest precedence. The evidence of catastrophic failure to adequately control mine ventilation is reflected in the 2980 explosions and 14 142 killed in the UK since 1850 and in the thirteen separate occasions when loss of life has exceeded the 167 fatalities of the Piper Alpha disaster.[15] Fortunately the loss of life from explosions in modern coal mines is much reduced. Ventilation schemes are in place with surface fans up to $400 \, m^3 \, s^{-1}$ capacity, assisted by underground booster fans. Colliery ventilation energy costs often exceed £0.5M yr^{-1}. Surface and cross-measure leakages can result in less than 40% of the surface fan airflow being effectively used for the ventilation of working districts.

In order to permit high face-production levels, of the order of 20 000–40 000 tonnes week^{-1}, specific measures are required to ensure methane emissions do not excessively constrain production limits. These include pre-drainage of coal seam gas, introduction of sewer gates, sequential working at short coal-faces, and three-road retreat working.[16] Future microbiological methane control techniques using strata infusion prior to mining and the establishment of an active 'biofilter' in the coal-face gas fringe zone could, in concept, reduce methane emissions by about 50%.

It is increasingly recognized that the coal seam methane content represents a valuable fuel resource and abstraction of methane takes place by two mechanisms:

[15] G. E. Green and H. C. Evans, 'Explosions, Ignitions, Rescue', Proceedings of the IME Symposium 'Safety, Hygiene and Health in Mining', Harrogate, UK, 18–20 November 1992, pp. 143–161.

[16] D. P. Creedy, 'Reducing Gas Emission Constraints on High-Production Retreat Coalfaces', *Min. Eng.*, June 1992, 335–338.

(1) in-seam, via boreholes drilled into, above, and below the seam horizon; and (2) via surface abstraction through a network of vertical boreholes with appropriate stimulation techniques.

Worldwide coal-bed methane resource base is estimated at 65 Tm^3.

Utilization of drained mines gas is being widely considered, primarily for cogeneration applications of 1 MW electrical generation and upwards. British Coal has undertaken several recent site specific assessments.[17] Both reciprocating engine and gas turbine utilization technologies have been used. Combined cycle gas turbine operation is favoured in larger generation schemes because of the higher thermodynamic efficiency ($\sim$45%) and high system availability, giving high financial rates of return. In view of the detonation risks during compression (the upper explosive limit of methane/oxygen mixtures rises with pressure), gas supplies must have a minimum methane content of 40%.

Environmental Monitoring

A range of fixed-point and hand-held instruments are available in a form approved for underground use to measure methane, carbon monoxide, oxygen, products of combustion, temperature, pressure, and air velocity. The sensor technologies utilized for gas sensing are generally of the electrochemical (O_2, CO, H_2, NO, NO_2), semiconductor (CO, products of combustion, *e.g.* Taguchi cell), or catalytic (CH_4, alkanes, *e.g.* Pellistor cell) types. These technologies serve as the basis of current monitoring schemes. Further work has concentrated on developing sensor techniques which are technically more robust, offering remote measurement, greater specificity, minimal cross-sensitivity, and reduced maintenance requirements. Prime examples would be the developments in optical methanometry. Embryonic gas sensor technologies using Langmuir–Blodgett films and enzymatic response are also noted.

Several sensor technologies have been assessed for fire detection purposes and refinements continue to provide discrimination against diesel fumes and to optimize operational deployment. For the detection of spontaneous combustion, tube bundle systems presently offer the most reliable method. A tube bundle system brings air from inbye locations underground via a network of small bore tubes to the surface for automated sequential sampling, recording, and alarm discrimination. The transit delay of the sample, approximately 20 min km^{-1}, is not a disadvantage for spontaneous combustion detection. Distributed optical fibre temperature sensing has been assessed in underground tests and offers the basis of future sophisticated spontaneous combustion detection and location systems.

Intelligent discriminating alarm algorithms are finding increased use to reduce false alarm rate.[18] The development of computer based mine monitoring systems is exemplified by the UK MINOS system.

---

[17] J.S. Moorhouse, 'Mines Gas Fired Power Plants', *Min. Technol.*, April 1992, 99–101.

[18] S.R. Hunneyball, *Min. Res. Eng.*, 1988, **1** (2), 115–124.

4 Mining Environmental Impacts and Controls

Previous discussion has concentrated on coal mining 'internal' or workplace environmental concerns. The principal external impacts and controls associated with coal production are now reviewed.

Opencast Mining Environmental Impact Control

The UK opencast coal mining industry is subject to a variety of social, geographic, and environmental operational controls. The average coaling duration of UK opencast sites is six years and sustainability of output is critically dependent on receiving planning consents for replacement capacity. To maintain output, twelve new sites need to be commissioned each year. Without further planning consents, output is reduced by 50% within three years. Geographically, the central shallow coalfields are juxtaposed to major conurbations. Consequently, UK opencast sites are often small, with boundaries delineated by habitation proximity, avoidance of infrastructure and environmental features, rather than by techno-economic considerations. Compounding matters, a major shift in public perceptions and attitudes have taken place over the last decade against heavy extractive industry developments.

The above factors act in unison, continuously challenging management. In response, British Coal Opencast integrates operations, environmental control, and restoration into a single coherent process. Environmental impact assessments are produced for every application and three principles are axiomatic to the integrated environmental management approach: (1) 'Polluter Pays Principle': the operator will meet the full costs of compliance; (2) 'Precautionary Principle': the onus is on the operator to demonstrate that environmental damage will not ensue; additional measures are required where there is uncertainty; and (3) 'BATNEEC': operator will use 'best available technique not entailing excessive cost'. Formalization of this approach is in progress.

A planning consent will typically contain 70 enforceable conditions. The major environmental impacts which must be addressed are: visual intrusion, dust control, noise, blasting, traffic, and hydrology. Preventative or mitigative approaches to these impacts are set out in Table 2.

Land restoration policies are becoming increasingly sophisticated, offering diverse conservation features, and a high resulting amenity value. The latter is especially relevant in that 25% of land granted planning permission for opencasting in the UK is classified as derelict and that 16 000 acres have been restored to date. The costs of restoration can represent a significant proportion of production costs. In the United States, the Surface Mining Control and Reclamation Act of 1977 (SMCRA) dictated that surface-mined lands be restored to support pre-existing or higher uses, that approximate original contours be restored, and that the disturbance to ground and surface waters be minimized. Costs of compliance with SMCRA vary, with estimates from \$1.90 to \$18 tonne^{-1} for Appalachian operators and \$0.10 to \$2.10 tonne^{-1} for western mines.

It is evident from the aforegoing discussion that environmental issues will remain the primary focus of planning and operational controls. British Coal

Table 2 Impact factors: standards and procedures[19]

Visual Intrusion	Embankment constructed from initial topsoil and subsoil: embankments 'landscaped' with relatively flat slopes and gentle curves; where possible, shrubs and trees established as part of ultimate restoration. Overburden mounds designed using perspective modelling. Plant yards, offices, *etc.*, screened.
Dust Control	Soil embankments grassed, overburden mounds increasingly hydroseeded. Internal haul roads continually sprayed by mobile water bowsers. Experiments with mobile vapour masts.
Noise	Baffle embankments, exhaust silencers on all internal combustion engines, electric plant where appropriate, acoustic enclosures fitted to fixed items of plant, restricted working hours. Regular monitoring.
Blasting	Limited to maximum peak particle velocity of $12\ \mathrm{mm\ s^{-1}}$.
Traffic	Where possible, internal haul roads and rail disposal. Where lorries use public roads, routes agreed with Planning Authorities: in some cases road improvements made at our expense. Lorries wheel-washed and sheeted on site. Timings agreed with Planning Authorities.
Water	Settlement lagoons to treat surface and pumped water which is directed through a system of cut off channels located around the excavation area and adjacent to the toe of mounds within the site boundary.

acknowledges this and supports a research programme to reduce the environmental impact of its surface mining operations. An example of this research is given in the following section on fugitive dust.

Fugitive Dust. Community perception of dust nuisance is complex, probably linked to threshold value changes accounting for pre-existing air quality and sociological characteristics of the subject area. Degree of nuisance has also been expressed in terms of visible contrast with and without clean reference surfaces. The situation is complicated in that there are no UK statutory levels of dust deposition that constitute an official nuisance. Other national dust deposition standards are inconsistent, with limit values between 130 and $650\ \mathrm{mg\ m^{-2}\ day^{-1}}$. General air quality criteria for the protection of public health, expressed in terms of suspended particulate matter concentration, have been defined by the US, EPA, WHO, and other authorities. These long-term guidelines are not, however, appropriate for designing short-term dust control responses. Equally, occupational exposure limits for particulates in air set by the UK HSE, whilst important for workers at mineral operations, do not relate well to nuisance perception. Gaussian plume, dust dispersion modelling is of value in predicting deposition rates at receptor points about 1 km and further downwind, yielding errors of 20% in annualized prediction data. However, predictions of deposition rates in the

[19] R. Proctor, 'Opencast Mining and the Environment', Proceedings of the World Mining Congress, XV, Madrid, 25–29 May 1992, Vol. II, pp. 1101–1107.

immediate vicinity of the mine curtilage can be overestimated by a factor of 2–3.

British Coal's strategy has been to institute comprehensive dust measurement surveys at active opencast workings and deep mine surface sites. Site-source and cross-boundary fluxes have been separately determined and correlated with a wide range of meteorological parameters which include wind speed, temperature, barometric pressure, solar radiation, rainfall, and soil moisture. By having a network of stations it has even been possible to track and quantify the source contribution from inter-continental transport of particulate. The exercise concluded that mean *respirable* dust concentration averaged for all sites was around $40 \mu g \, m^{-3}$ and that on-site concentrations were below $1.5 \, mg \, m^{-3}$, well below occupational dust exposure standards. There was no evidence of significant on-site emissions from colliery surface operations.

This work is an important step towards an optimized dust sampling strategy and implementing effective on-site controls predicated on meterological measures allied with good design practice. This should reduce further the already small health risk to the workforce and the incidence of public complaints, which at most sites averages less than one complaint per year. Work continues to examine performance enhancements and simplifications of traditional methods of fugitive dust measurement, including relative directional dust gauges. Note is also made of the extensive variety of mechanical and chemical dust control technologies which have been developed.

Deep Mining Environmental Impacts and Controls

Colliery Spoil Disposal.   The problem of colliery spoil disposal was a central term of reference of the UK Commission on Energy and the Environment by the Secretaries of State for Energy and the Environment which produced the Flowers Report in 1981. The Government's response to this report confirmed its commitment to the principle that the polluter pays.

The implications for production costs are considerable. Assuming a coal to dirt ratio of 2:1 for the Belvoir coalfield, spoil generation at Asfordby could exceed $600\,000$ tonnes yr^{-1}. Submissions to the associated public inquiry identified the cost of spoil disposal to be £3.82 to £11.34 tonne^{-1} for available remote sites compared with £2.10 tonne^{-1} for tipping adjacent to the source.

About 80% of waste is dry of solid spoil, traditionally tipped close to the source. The remainder is wet spoil, a by-product of coal beneficiation. This is disposed of mainly by lagooning. The principal environmental effects of tipping are visual intrusion, loss of land, noise and dust during tipping, and potential water pollution. Tip combustion and structural failure are also hazards. Tip collapse is rare now that the mechanisms of failure are better understood.[20]

In response to the requirement to find acceptable spoil disposal avenues the following options are available: (1) reduce waste generation at source; (2) rework colliery tips for their fuel content; (3) use spoil as in-fill for large civil engineering projects; (4) process waste to produce secondary products; and (5) dispose of waste underground.

[20] R. N. Chowdhury, V. U. Nguyen, and J. A. Nemcik, *Min. Sci. Technol.*, 1986, **3**, 127–139.

Comments are made on each. Reducing waste generation at source, principally by the use of selective mining techniques, coal-stone batch discrimination, and face automation is considered in section 5. The use of dewatering technologies to increase fine coal utilization and thereby reduce lagoon disposal is noted.[21]

The disposal of coal washery tailings and lagoon fines by fluidized bed combustion, possibly with electricity generation, is feasible providing there is consistently adequate carbonaceous waste to burn autothermally. Coarse rejects and fine rejects (tailings) are the most likely feedstock in this case. Demonstration systems have been built in Australia. Given the high inert materials content, ash disposal needs to be carefully considered.

The use of minestone for road embankments and other large scale civil engineering projects is widely practised. The material compaction, compressibility, and shear strength characteristics have been shown to have acceptable stability. The geotechnical characteristics of coal mining wastes of various coal-producing countries are also broadly similar.

The development of secondary products from colliery spoil is unlikely to substantially mitigate the scale of the disposal problem. Greater use can, however, be made of coal combustion residues (fly ash, bottom ash, fluidized bed combustion ash, flue gas desulfurization sludge). Typical uses are cement, artificial gravel, concrete, and road base materials.

Disposing of waste underground, particularly by incorporation as a cyclical component of the longwall mining process, has been widely investigated.[22] In all underground waste stowing systems the question is not so much the requirement for further technological development but rather that of economic feasibility. Both pneumatic and hydraulic transport methods show considerable promise for handling large volumes of waste. However, the stowing process inevitably slows the advance rate of highly mechanized faces, lowering system utilization, and increasing costs. Reduced subsidence is noted as a benefit of the backfilling of cavities with waste.

Studies have also been conducted to use coal mines for storing other wastes, including radio-active residues. There are, however, fundamental associated geotechnical problems; namely, overburden water permeability is greatly increased by mining operations, and minewater ion content leads to high waste elution rates. For radio-active residues, rock salt formations are favoured because of the high plasticity, homogeneity, and thermal conductivity of the salt geology.

Subsidence. Ground movement is inextricably linked with most types of underground mining. Mining creates a void which may cause the overlying burden to subside, moving vertically and laterally into the excavated space. Movement continues until the bulking of the rock material has closed the space or the compressive forces have been placed again in a state of equilibrium. The surface influence of ground movement (the subsidence basin) is generally greater in area than that of the void underground.

[21] D. W. Brown, 'Turning Colliery Waste into Saleable Products', *Trans. Inst. Min. Metall.* (*Sect. A: Min. Ind.*), September–December 1990, **99**, A133–A137.

[22] R. Astle, 'The Underground Disposal of Mine Waste', *Mine Quarry*, January/February 1991, 70–71 and *Mine Quarry*, March 1991, 26–28.

The application of precise deformation measurement, development of empirical tools, and construction of models began in earnest after the nationalization of the UK mining industry.[23] Empirical, profile, and influence function methods of prediction have been complemented by various analytical models and physical and finite element models. There are errors associated with any subsidence prediction approach, typically between $\pm 10\%$ and $\pm 30\%$ at the centre of the subsidence basin. In any event, local conditions must be fully taken into account. Field observations show, for example, significant differences between competent strata of the Newcastle Coalfield in New South Wales, Australia, and typical British conditions.

Modern subsidence control uses the large body of regional empirical data to augment modelled prediction and controls. The net consequence is that whilst the mining engineer must live with the intrinsic association between extraction and subsidence, its adverse effects can be managed and largely mitigated.

Minewater. The majority of coal mines encounter groundwater during mining, from the mine development phase through to abandonment. Water creates environmental, operational, and stability problems. In surface mining safety hazards are associated with instability through various ground failure mechanisms. Alteration of river hydrology is also a serious issue. Underground mining safety concerns relate to large-scale rapid inflows from surface water bodies, aquifers, or previous workings above the mining zone. Certain regions, such as the Upper-Silesian coal basin, characterized by complex hydrogeology and low strength strata, have particular problems. To prevent groundwater flow in interconnected old workings towards working collieries requires very large pumping infrastructures. For example, pumping protection schemes in the East Durham and Nottinghamshire coalfields pump 36 billion litres and 14 billion litres per year, respectively.[24]

The restrictions on discharging minewaters to watercourses are becoming increasingly severe. The principal problems, in terms of the constituents of the water, can be summarized as follows: quantity involved; suspended solids; iron compounds; dissolved salts; ammonia; oxygen demand; and acidity.

Various treatment schemes are available, including chemical precipitation, reverse osmosis, ion exchange, flash distillation, and biological passivation.

The major focus of international concern of all the water management issues is that of acid mine drainage. The acidic nature and high associated sulfate, ferrous iron, and base metal content can contaminate groundwater and watercourses and damage local ecosystems. The mechanisms of oxidation and the mediating role of acidophilic bacteria are well understood. However, there is no definitive answer as to the most appropriate means of prevention, treatment, and control, particularly when acid generation may persist for many decades following mine closure. Trace element mobilization is also not fully understood. Chemical

---

[23] B. N. Whittaker and D. J. Reddish, 'Subsidence—Occurrence, Prediction and Control', Elsevier, 1989.

[24] P. L. Younger and G. Reeves, 'Protection of the Environment—Mitigation of Environmental Impact, With Particular Reference to the North East Coalfield', Institution of Water and Environment Management Symposium, 'The Environmental Effects of Coal Mine Closures', University of Nottingham, 1 April 1993.

treatment of acid mine drainage at source is technically feasible and models are available to support plant design. The mechanisms of acidic surface water neutralization are, however, less well understood. Inorganic bases may be administered, but ultimately a sustainable biological buffer is required. It may well be that future acid drainage treatment schemes will defray the costs of treatment by recovering the metallic elements as purified fractions. Recovery by fast solid phase extraction is one possible supporting technology.

Methane Emissions.   Protocols to reduce methane emissions are anticipated as the global warming debate advances. Emissions of methane associated with human activities account for $\sim 60\%$ of net global emissions of 540 Tg p.a. The global contribution from coal mining and specific emissions ('release factors') are a subject of scientific contention. Most studies cite values for global emissions from coal mining of between 30–50 Tg p.a.[25] Specific emissions vary according to a number of factors, including depth, pressure, and coal rank. The mean specific emission for the UK is $15.3\,m^3$ tonne^{-1}. Surface mines and abandoned mines have relatively small methane emission fluxes.

The control measures available to reduce emissions from the mining process include degasification by vertical wells, in-mine drainage, combustion of methane in the exhaust ventilation air, and microbiological coal-seam degasification. Methane gases from mining will increasingly be regarded as an environmental hazard and wasted resource. Those technologies associated with useful electricity generation will find increasing application. The methane emissions associated with mine closure and coal storage, though not large, do present gas accumulation hazards and must be carefully controlled.

5   Technology to Mitigate Environmental Impact

The previous sections of the paper have reviewed workplace (internal) and external impacts of coal extraction. This section develops three specific technological themes to mitigate environmental impact. The three examples chosen are appropriate to current/near-term, middle-term, and longer-term responses, respectively. Clearly there is an ascending element of speculation in the latter two cases.

Current/Near-Term Theme: *Quality Control and Automation*

Machine Steering.   Dirt reduction begins at the point of production. On longwall faces, optimized vertical steering is the key to a higher quality run-of-mine (ROM) product. Allied with this approach, in-seam drivages supported by roofbolts minimize dirt production during face development. Automatic machine steering brings the joint benefits of improvements in productivity, improvement in product quality by minimizing the dirt cut, and better strata control, with an associated reduction in machine downtime.

In essence machine steering is a means of automatically controlling the vertical

[25] IEA Coal Research, 'Methane Emissions from Coal', IEA Perspectives, IEAPER/04, November 1992.

position and height of the extracted section within a coal seam within required limits. If the vertical position or height of the extracted section within the seam has to be changed, then the steering system accomplishes this in a controlled manner, ensuring that the advancing face equipment negotiates the changing seam horizon without disruption of the mining process or unnecessary cutting of dirt.

There are consistent reports of increased production rates and reduced dirt by the application of horizon control systems.[26] Notwithstanding the large variance in the gains achieved, production rates increase by $\sim 20\%$ and dirt reduction is at least 10%. Where selective face cutting is initiated to avoid discrete dirt bands and high sulfur bands, then the dirt reduction factor can be much higher.

To accomplish auto-steering requires a multiplicity of associated sensors to determine machine/drum/seam/previous extraction spatial relationships and along-face location. The central critical technology is the coal seam interface detector. There are two generic forms of coal interface detection system: coal thickness measurement and coal–rock interface measurement. The former consist of nucleonic or electromagnetic sensors that gauge some type of signal penetration through a desired thickness of coal. The latter generally consist of vibration, infrared, and optical/video sensors which require actual contact or visibility of the interface. In the UK, coal thickness sensors are required since it is normal practice to leave a layer of roofcoal to help maintain integrity of the immediate roof strata.

Early experiences pointed to roof guidance systems being simpler to implement than seam floor guidance. Subsequent interface detector developments concentrated on roof coal interface detection. The first UK roof guidance systems used a synthetic radiation source (Caesium 137) mounted within a probe which emitted gamma rays into the roof coal. The detector measured the intensity of the backscattered radiation. This device was rapidly superseded by the invention of the natural gamma coal thickness measurement sensor.

The operational requirement for the natural gamma sensor is that there is a useful ratio between the gamma activity of coal and the immediate roof, which usually consists mainly of shaley materials. The US Bureau of Mines has conducted a wideranging survey of US coal seams and has determined that 81% of underground mines have an immediate roof composed of shaley materials (*e.g.* shale, draw slate, clay, and claystone), *i.e.* potentially suitable for natural gamma application. Where natural gamma is inapplicable (*i.e.* low gamma contrast or total removal of roof coal is required) then other interface detection schemes are required, but at this time there is, unfortunately, no technically robust alternative available.

The Integrated Face. Control and monitoring systems for the principal coalface sub-systems have largely been developed separately, leading to a situation where 'islands of automation' exist on the coalface, each improving the efficiency of a specific operation but working in isolation from the others.

British Coal's strategic response has been to direct the development of fully integrated sub-systems. The objective is to significantly increase the utilization of

[26] D. J. Buchanan, R. Astle, J. Emery, T. F. Jones, and P. M. Taylor, 'Reducing the Environmental Impact of Dirt from Underground Mining Operations', *Min. Eng.*, May 1992, 315–318.

88

longwall faces which currently operate at around 40% of their potential. Automation of the overall production cycle should permit increased equipment utilization, improved product quality, allow quasi-continuous operation, and reduce manpower requirements and hazards.[27] The sub-systems required for the automation process are as follows:

(1) Vertical horizon control of the coal-cutting machine.
(2) Automatic advance control of the coalface.
(3) Electronic initiation and automatic sequence control of the face support system.
(4) Control of the haulage system of the power loader to relate cutting speed to load on the AFC motors.
(5) On-line condition monitoring.
(6) Safety monitoring and personal protection.
(7) Face control centre linking local sub-systems and to surface.
(8) A surface supervisory system.

Development of the UK Integrated Face programme is a collaborative project, part funded by the DTI involving Anderson Longwall, Meco (now Longwall International), NEI, Leeds University, and British Coal.

Improving Product Quality. The application of automated coal shearer horizon control will materially improve the quality of the raw mined coal. In spite of this improvement a significant dirt component will remain, often intimately bound with the product. The raw mined coal is rarely suitable for direct shipment to the principal customers, the generating stations. For electricity generation, the saleable product is usually obtained by blending washed coal with a proportion of untreated coal to achieve the target specifications. The specifications for the quality and consistency of the delivered product are demanding, with large penalties for inconsistencies in net calorific value, ash, moisture, sulfur, chlorine contents, or inadequate handleability.

To meet the operational and environmental quality demands of the customer, British Coal and other coal producers are in the process of adopting dynamic process management systems which use sampling and rapid off-line analysis or on-line ash, moisture, and elemental analysis monitors to provide feedback in the control scheme, frequently based on statistical process control techniques.[28] An outline of the control strategy used by British Coal is shown schematically in Figure 4. This quality management system makes use of computer-based, predictive statistical models. Underlying product trends are derived and related to product price structure and specific penalty clauses to provide guidance on changes to plant operating parameters necessary to reduce coal quality variances and optimize proceeds.

To increase clean coal recovery and reduce wastage, it is also necessary to have a washing plant which is efficient, controllable, flexible, and reliable. The process technologies associated with coal cleaning are extensive, and techno-economic

---

[27] A. T. Shaw, 'Moving Towards the Integrated Face', *Colliery Guardian*, September 1992, 192–196.
[28] D. J. Buchanan and T. F. Jones, 'Why Quality Coal Demands Integrated Preparation', *Colliery Guardian*, September 1992, 187–191.

Figure 4 Integrated coal
preparation

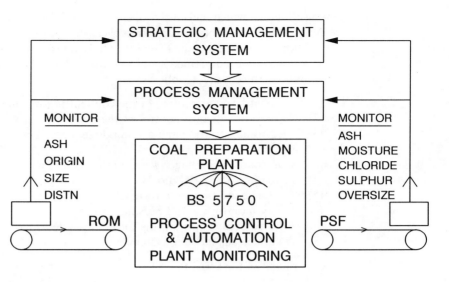

judgements of the merits of particular processes are complex. British Coal views
the novel dense medium separator configuration which it has developed
(LARCODEMS—Large Coal Dense Medium Separator) as a fundamental
technology in future low cost flexible automated coal preparation plants. Figure
5 illustrates the novel hydrodynamic operation of the two LARCODEMS variants.

Figure 5a Two product
'LARCODEMS' separator

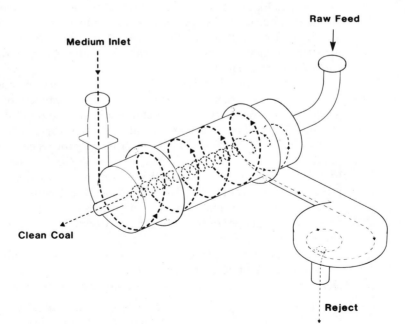

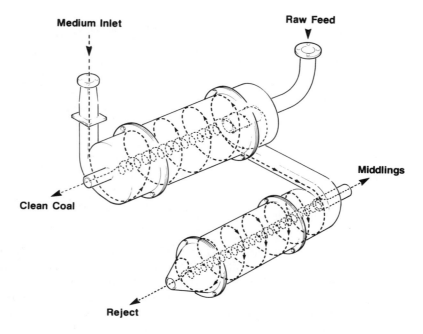

Figure 5b Three product 'LARCODEMS' separator

Medium-term Theme: *Fuel Cycle Integration*

The medium-term theme elaborates a basic premise that, from a least cost planning approach, there is scope for increased vertical integration and synergy in the coal extraction–utilization cycle, primarily via a coal–water fuel approach. This premise is speculative and its advancement is subject to significant progress being made in the effectiveness and costs of advanced coal cleaning. These developments are seen as medium-term. In the near-term, fuel switching to gas and low sulfur coals will further inhibit adoption.

Figure 6 shows a schematic representation of the 'coal-to-electricity' chain. The established approach is for coal to be mined, washed, and transported to the power station. For conventional pulverized fuel thermal stations the fuel must meet tight quality criteria and must have good handling characteristics, dictating a low delivered moisture content. Excessive fines component in the mined coal is a major problem. The costs of cleaning and then drying fine coal are considerable. This restricts the application of advanced beneficiation processes, which fundamentally require a micronized product to be effective. Thus current market demands for a dry, easily handled solid product limit scope for enhanced ash, sulfur, and chlorine removal at the beneficiation stage.

At the power station (assuming conventional large pulverized fuel plant) the coal is milled and combusted. Any emission control measures are, in the main, 'end of pipe' technologies. Retrofitting of flue gas desulfurization equipment is frequently viewed as a 'negative investment' by utilities, and net station thermal efficiency is lowered. Added to this, the likely emergence of critical load approaches to SO_x and NO_x control may require moderate reductions in the

Figure 6 The
'coal-to-electricity' chain

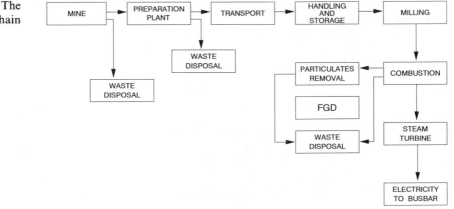

emissions of a number of stations. The control strategy of using FGD on a small number of large stations, previously applied to meet national 'bubble' emission limits, may be inappropriate under these circumstances.

The response option advanced is for mining/fuel companies to progress towards providing a deep cleaned, micronized product with intrinsically higher value, and environmental acceptability. This could, in principal, be provided as a dry product or as a slurry. There are several reasons for favouring the latter wet-processing route. The generators would be provided with a product with simplified bulk handling, storage, fly ash disposal, and flue gas cleaning requirements. Associated preparation costs would be defrayed against reduced generating operating costs, and avoidance of FGD retrofit/boiler replanting costs. Fuel would be delivered as a pseudo-fluid, permitting pipeline transport from mine-mouth fuel preparation plants, possibly with a 'just-in-time' supply philosophy. The supporting rationale for the utilities and mining companies, respectively, can be summarized as follows:

Utilities—

extension of PF station economic life; permits concentration on core generation activities; mitigates non-base load operating costs; addresses critical loads requirement for station-by-station emissions reduction; shifts capital investment requirement more to fuel supplier; simplified handling and transportation; reduced waste inventory; lower environmental external costs.

Mining Companies—

value added, environmentally enhanced product; scope to enhance competitiveness and market security by using advanced process technologies; delivered product price has reduced sensitivity to mining cost component; philosophy encourages 'fine coal-wet circuit' approach. Totally inverts traditional fine coal problems; pipeline transportation option.

For coal–water mixtures (CWM) to advance to large scale demonstration in Europe it is anticipated that EC support will be required, possibly via the Thermie programme. The trend for utilities to exercise differential purchasing policies, which discriminate against high sulfur and ash content coals, may assist the introduction of processed (coal) fuels. Relative progress in integrated gasification and fluidized bed technologies will also have an effect on the choice of technologies.

The economies of CWM fuel cycles are improving. Prior to 1985 the prospects for CWM projected a modest global market potential of 60 Mt yr^{-1}, mainly as a residual fuel oil substitute. More recent cost studies suggest that CWM is already competitive with heavy fuel oil in Japan, and that for large annual plant production levels ($>$Mt yr^{-1}), cost is comparable with pulverized coal use. Initial application favours long distance overland hydraulic transport. There are several CWM demonstration plants, with greatest interest in Japan, Italy, and Canada. Some prospective commercial CWM users are progressing via an intermediate step of evaluating combustion of coal–oil–water mixtures.

A breakout of cost contributions associated with the utilization of Australian coal at a modern Japanese power station highlights the scope to reduce power station costs:

Mining	1.0–1.5 US cents (kW h)$^{-1}$
Coal Preparation	0.1–0.2 US cents (kW h)$^{-1}$
Rail/Port/Ocean Freight	0.6–0.8 US cents (kW h)$^{-1}$
Milling/Combustion/Generation	3.0–3.5 US cents (kW h)$^{-1}$
Emissions Control	1.3–1.6 US cents (kW h)$^{-1}$

The potential benefits of shipping a 'refined', high value added product to utilities has not been lost on Australian producers.[29]

It is difficult to estimate which beneficiation process(es) will predominantly be used to produce low ash, low sulfur products in the future. Advanced physical cleaning appears more likely than either chemical or biological processing. Promising technologies include the micronized magnetite dense medium route and the two-stage oil agglomeration route. Certain secondary effects will also have to be taken into account, such as trace element pathways. Comparative costs for existing and developing categories of advanced coal cleaning have been reviewed by Yeager.[30]

Longer-term Theme: In situ Mining/Conversion

The final theme deals with long-term, 'over-the-horizon' extractive technologies. Advocating which of the present day embryonic or emerging research technologies will advance to mature technology status in the first quarter of the next century involves intelligent conjecture.[31]

Historical reviews of mining and the role of technology conclude that mining has evolved incrementally with increasingly more productive mechanical systems substituting for human labour or less productive machines. A cautious projection

[29] A. Broome, 'Green Coal: Markets and Technology', Abridged presentation to Australian Coal Conference, May 1992 in 'World Mining Equipment', November 1992, 20–24.

[30] K. E. Yeager, Energy World, June 1990, **179** (suppl.), 2–14.

[31] K. Moses, 'Technology in the Next 100 Years—Revolution or Evolution', Min. Eng., September 1989, 115–118.

would infer that further incremental improvements will continue to take place in all mining system components. The underlying drivers for development will be demands for lower production costs realized through higher system productive capacity or quasi-continuous operation, reducing hazards to the workforce, and the mitigation of the negative environmental aspects of mining.

The US Bureau of Mine's view of future systems anticipates the removal of the workforce from the hazardous face area. Autonomous or teleoperated plant will exploit advances in robotic and artificial intelligence enabling technologies. Continuous miners have been targeted in research work with navigational, guidance, diagnostic, and coal interface sensor techniques currently being appraised. This could determine the basis of future US mining technology.

Incremental change is the dominant form of technology evolution in mining. Radical breakthroughs which render existing plants and processes obsolete or enable uneconomical mines to become viable are rare in the mining industry. Notwithstanding this, there are several embryonic technologies which may form the future technological basis of mining. It is the **energy content** of the coal rather than the coal itself which is sought, and therefore *in situ* conversion and remote recovery are potential options. Coal will also find a role in the longer-term as a petrochemical industry feedstock and for the synthesis of liquid fuels. However, this view is tempered by the potential adverse environmental impacts of large-scale coal liquefaction.

Alternative non-conventional extraction techniques broadly comprise the following: underground coal gasification; complete combustion underground; pyrolysis; quenched combustion to produce high temperature steam; solvent digesting using a coal derived oil; aqueous phase liquefaction using high temperature water; supercritical gas extraction pyrolysis; chemical comminution to produce a coal slurry; microbiological degradation to produce methane; microbiological solubilization of coal; hydraulic mining; liberation of coal-bed methane. Of these options, development resources are essentially concentrated on underground coal gasification and surface extraction of coal-bed methane.

Underground coal gasification (UCG) is an intriguing technology. In the process coal is ignited in place, producing heat that pyrolyses and gasifies further coal. Controlled air/oxygen and steam injection results in a product gas stream of medium calorific value. The gases are injected and retrieved using vertical well technology. In principle, the *in situ* conversion, utilization of currently unmineable resources, reduction in coal waste by-products, and potentially lower capital costs are key UCG advantages over conventional mining.

The development of UCG has been long in gestation. UCG was first theorized by Siemens in the late 1860s. The first patent for UCG was granted in 1909 in Britain. Early work commenced in 1913, but initial aspirations were not realized. Systematic UCG research began during the 1930s and numerous national programmes have been conducted since. The European Community is at this moment engaged in pre-commercial trials of a deep UCG plant in Spain.[32]

---

[32] M. A. Albeniz, J. M. Gonzales, A. Obis, E. Menendez, V. Chandelle, M. Mostade, and A. C. Bailey, 'Joint European Underground Coal Gasification Project in Spain', Proceedings of the World Mining Congress, XV, Madrid, 25–29 May 1992, Vol. I, 291–300.

6 Conclusions

An attempt has been made to define the complex range of issues which confront the coal mining industry today: competitive sourcing, open markets, workplace and surface environmental controls, globalized environmental legislation, and response strategies.

An emergent view is that the coal resource, in all its phases from exploration through to mining and subsequent utilization must be viewed in a holistic manner and managed within a comprehensive environmental framework covering all cycle phases. Health and safety of the workforce must equally be accorded high priority. The embrace of mining methods and technologies which are compatible with sustainable development will consequently be a prerequisite to secure a long-term role for coal. The costs of environmental protection may appear increasingly demanding, potentially even reducing competitiveness in the short-term. However, it is a challenge to which the technologically advanced mining nations must rise. For this, research must continue to be directed, imaginatively and effectively, at the spectrum of environmental responsibilities. Specific examples have been cited. The global environment will then be the main beneficiary of transferring the ways and means to ensure that coal is mined and used efficiently and sustainably.

Acknowledgements

The views expressed in this paper are those of the authors alone and do not necessarily represent those of the British Coal Corporation. The assistance of many colleagues in the preparation of this paper is gratefully acknowledged.

Methane Emissions from Coal Mining

A. WILLIAMS AND C. MITCHELL

1 Introduction

The importance of methane (CH_4) as a greenhouse gas is increasingly being recognized and has been the subject of numerous reviews.[1-4] There are many sources of methane, usually involving the degradation of organic matter to a more thermodynamically stable form, namely methane—and these include sources such as biomass, landfill, petroleum, and coal.

The coal industry is of considerable size and is spread over fifty countries. The total coal production worldwide in 1991, including brown coal and lignite, according to IEA Coal Research, was 4566 Mt. The major producers of hard coal in 1991 were China, 1087 Mt; USA, 825 Mt; former USSR, 485 Mt; India, 229 Mt; S. Africa, 181 Mt; Australia, 167 Mt; and the UK 96 Mt, whilst other countries are major producers of brown coal and lignites Colombia and Indonesia are becoming major producers. It is difficult to predict the role of coal over the next few decades because of the competing influences of natural gas and nuclear energy. At present 40% of the world coal is used for electricity production and, whilst there may be an initial decrease in coal use (at least in Europe, because of the increased use of natural gas), it seems likely that after a few years there will be a worldwide increase. IEA Coal Research suggest coal consumption will expand to 5500 Mt by the year 2000. Clearly methane emissions from coal production, handling, and combustion will be a significant component of global methane emissions for a considerable time.

The geological formation of coal, commonly called coalification, which involves an increase in carbon content and density as it changes from peat via lignite (65–72% mass C) to the hard coals such as bituminous coal (76–90% mass C) and anthracite (93% mass C), results in methane formation together with carbon dioxide and nitrogen as shown in Figure 1.

The quantities formed can be substantial as shown in Figure 2 although the

[1] T. M. L. Wigley and S. C. B. Raper, *Nature* (*London*), 1992, **357**, 293.
[2] Estimation of Greenhouse Gas Emissions and Sinks. Final Report OECD Experts Meeting, 18–21 February 1991. Prepared for Intergovernmental Panel on Climate Change. Revised August 1991.
[3] Methane Emissions and Opportunities for Control, Workshop Results of Intergovernmental Panel on Climate Change, September 1990.
[4] International Workshop on Methane Emissions from Natural Gas Systems, Coal Mining and Waste Management Systems, 9–13 April 1990, Washington, DC, USA.

Figure 1 Gas quantities generated during coalification. Reprinted with permission from Reference 5.

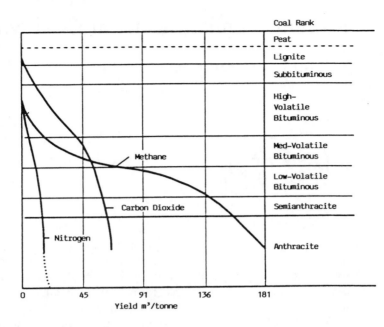

actual quantities found today are a function of earth movement and erosion. They are, as might be expected, a function of depth, pressure, moisture content, and the extent of coalification, and have been discussed by a number of authors generally relating to specific countries or areas such as the US, Australia, Germany, and the UK.[4,10] Typical trends for US coals are shown[5] in Figure 2. Coal rank represents the differences in the stages of coal formation and is dependent on the pressure and temperature of the coal seam; high rank coal, such as bituminous coals, contain more CH_4 than low rank coal, such as lignite. Depth is important because it affects the pressure and temperature of the coal seam, which in turn determine how much CH_4 is generated during coal formation. If two coal seams have the same rank, the deeper seam will hold larger amounts of CH_4 because the pressure is greater at lower depths, all other things being equal. The translation of the methane content of coal into methane gas emission during mining is a complex matter involving the nature of the storage of the methane in the coal, its transmission through coal beds during mining operations, and the mining operations themselves where degassing and ventilation practices may vary.

This complexity fundamentally arises because the methane exists within the

5   M. C. Irani, P. W. Jeran, and M. Deul, 1974, ICF Resources, 1990.
6   D. P. Creedy, *Q. J. Eng. Geol.*, 1991, **24**, 109.
7   D. Buchanan and D. P. Creedy, Report produced for the Working Group on Methane Emissions, The Watt Committee on Energy, 1993.
8   C. Mitchell, Energy Policy, 1991, 849.
9   I. M. Smith and L. L. Sloss, Methane emissions from coal, IEAPER/04, IEA Coal Research, 1992.
10 Coal Industry Advisory Board, Global methane emissions from the coal industry, October, 1992.

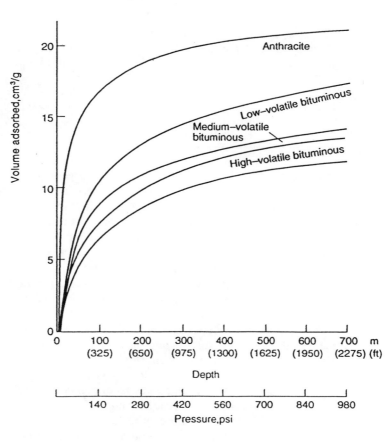

Figure 2 Relationship between adsorbed methane volumes and depth and pressure for different coal ranks. Reprinted with permission from Reference 5.

micro-pore structure of the coal, mainly as adsorbed layers. The amount present is effectively a function of the available internal pore surface area—a function of coalification—and the level of competing adsorbing or pore blocking species, such as water, and is, of course, a function of pressure.

2 Knowledge About the Emission Sources

Emissions may occur from operational and abandoned deep mines and from surface mines which may operate at a variety of depths and which have a relatively large exposed area. In addition, there are emissions from mined coal during transportation and combustion, but the latter are not dealt with here.

Emissions from Deep Mining

The calculation of methane emissions from coal mining is complex. Methane emissions and the rates of release will differ at each mine due to the way the coal is mined, the different qualities of coal mined, the different working depths, the types of ventilation, the different drainage or predrainage and capture systems

Table 1 Estimated underground emission factors for selected countries

Country	Emissions Factor/m³ tonne⁻¹
Former Soviet Union	17.8–22.2
United States	11.0–15.3
Germany (East and West)	22.4
United Kingdom	15.1
Poland	6.8–12.0
Czechoslovakia	23.9
Australia	15.6

that are used, and other factors such as the geology of the mining area; for example, the density of coal seams. In most underground mines methane is removed by ventilation in which large quantities of air pass through the mine and this air, typically containing a concentration of one per cent methane or less, is exhausted into the atmosphere. In some mines, however, more advanced methane recovery systems may be used to supplement the ventilation systems and ensure mine safety. These recovery systems involve drilling into the unmined coal seams and typically produce a higher concentration of methane ranging from 35% to 95%. In some countries, some of this recovered methane is used as an energy source, while other countries vent it to the atmosphere. Recent technological innovations are increasing the amount of methane that can be recovered during coal mining and the options available to use it. Thus, methane emissions could be reduced from this source in the future.

In an earlier OECD methodology,[4] a single emission factor of 27.1 m³ of methane tonne⁻¹ of coal mined was recommended for all underground mining. This factor included both emissions from mining and from post-mining emissions associated with underground coal production.

Based on more recent studies and additional country-specific emission data, the 1993 meeting of the OECD Experts Group in Amersfoort, The Netherlands recommended revising this emission factor to reflect some additional issues.[11] First, use of a range of emission factors is suggested to reflect the large variation possible in methane emissions from underground mines in different coal basins and countries. Second, this emission factor should represent only those emissions associated with underground mining and post-mining emissions should be handled separately.

The OECD Experts Group[11] recommends revised global average emission factors of 10 to 25 m³ tonne⁻¹ of coal mined (not including post-mining activities). This range reflects the findings of various country studies, as shown in Table 1. As more detailed emissions data are published by various countries, the factors can be further revised, if necessary.

In order to calculate methane emissions from coal mining, very detailed mine-by-mine, seam-by-seam validated information is required; however, in some smaller coal producing countries a simpler approach may be used. Therefore a three tier approach was devised.

The first approach—called the Global Average Method—uses a pre-determined

[11] D. Kruger, *Proceedings of the International IPCC Workshop on Methane and Nitrous Oxide*, Amersfort, Ed. Ar. van Amstel, RIVM, 1993.

range of emission factors (based on experience in a number of countries) to estimate emissions. The most complex third approach—called the Mine Specific Method—develops emissions estimates using detailed emission data for most, if not all, of a country's underground coal mines. In between these two methods is an intermediate second approach, called the Basin or Country Average Method, in which more limited information, including either measurements from a subset of mines, can be used to refine the range of possible emission factors presented in the Global Average Method. Each of these approaches is described in more detail below.

Tier 1: *Global Average Method.* The simplest method for estimating methane emissions is to multiply the underground coal production by a factor or range of factors representing global average emissions from underground mining, including both ventilation and degasification system emissions. This method may be selected in cases where total coal production from underground mines is available but more detailed data on mining emissions, geological conditions, and coal characteristics, are not; thus there is a high degree of uncertainty associated with it. The Tier 1 Equation is shown below.

$$\text{Low CH}_4 \text{ Emissions (tonnes)} = \text{Low CH}_4 \text{ Emission Factor [m}^3 \text{ CH}_4 \text{ (tonne of coal mined)}^{-1}] \times (\text{Underground Coal Production (tonnes)} \times \text{Conversion Factor}$$
(1a)

$$\text{High CH}_4 \text{ Emissions (tonnes)} = \text{High CH}_4 \text{ Emission Factor [m}^3 \text{ CH}_4 \text{ (tonne of coal mined)}^{-1}] \times \text{Underground Coal Production (tonnes)} \times \text{Conversion Factor}$$
(1b)

Where the Low CH_4 Emission Factor equals $10 \, m^3 \, tonne^{-1}$ and the High CH_4 Emission Factor is $25 \, m^3 \, tonne^{-1}$. Note the Conversion Factor converts the volume of CH_4 to a weight measure based on the density of methane, $1.49 \times 10^9 \, m^3 \, Mt^{-1}$.

Tier 2: *Country or Basin Average Method.* The Tier 2 approach, which is a Basin or Country Average Method, can be used to refine the range of emission factors used for underground mining by incorporating some additional country- or basin-specific information. Basically, this method enables a country with limited available data to determine where within the range of global average emissions their underground coal mines are likely to lie. Some countries, for example, may have enough available data to determine that their mines are quite gassy and that the low end of the global average range significantly underestimates their emissions, while other countries may find the opposite.

The best means of making this assessment is to examine measurement data from a limited number of underground coal mines to estimate where within the Global Average Emission Factor range a country's mines fall and what a reasonable narrower range of emission factors might be. Making this estimate will require judgement on the part of the estimator regarding the adequacy of the available data and its uncertainty. If sufficient expertise is not available to make such judgements, it is recommended that the Tier 1 approach (the Global Average Method) be used to prepare emissions estimates.

In some cases, measurement data on emissions from mines may be unavailable but a country will still seek to develop a more refined estimate based on other types of available data. In such cases, a country may seek to develop a simple emissions model based on physical principles or make judgements based on an evaluation of available data.

It should be noted that while the Tier 2 approach can provide some additional information about methane emissions in a particular country or coal basin, the estimates will still be quite uncertain because of the absence of comprehensive and reliable emissions data. This approach should thus be used only in cases where there is a strong need to make an estimate that is more refined than the Tier 1: Global Average Method, and not enough data are available to prepare an estimate using the Tier 3: Mine Specific Method.

Tier 3: Mine Specific Method. Because methane is a serious safety hazard in underground mines, many countries have collected data on methane emissions from mine ventilation systems, and some also collect data on methane emissions from mine degasification systems. Where such data are available, the more detailed Tier 3 approach, called the Mine Specific Method, should provide the most accurate estimate of methane emissions from underground mines. Since these data have been collected for safety, not environmental reasons, however, it is necessary to ensure that they account for total emissions from coal mines. The key issues that should be considered when using mine safety data, as well as the recommendations of the OECD Experts Group for resolving them, are shown in Table 2.

The only data required for the equation is the *in situ* methane content of a seam, or an average *in situ* methane content for a mine or country. However, because of the factors outlined, the total quantity of methane emitted from the mine significantly exceeds the *in situ* methane content of the mined coal, this factor being typically from 2–5 in Europe and 4–10 in the USA. The relationship between adsorbed methane volumes and depth and pressure vary for different coal ranks as shown in Figure 2.

These two basic approaches that can be adopted for the estimation of the contribution of methane as a result of the mining of coal are discussed further in the following sections.

The Technique Based on In situ *Methane Content*

Here the methane emissions are related to the *in situ* methane content as described above and this method had earlier been suggested for adoption for the IPCC/OECD inventory.[2] The emission equation selected is:

Methane emissions in m^3 of methane = (2.04 × *in situ* methane content) + 8.16
per tonne of coal mined

This emission equation was derived from a detailed, empirical analysis[5] of the measured total methane emission rates from 50 US mines. It was found, when using the standard deviation around the mean as a measure of uncertainty, that the actual methane emission value could be up to 23% higher or lower than the

Table 2 Key issues for consideration when using Tier 3: Mine Specific Method

Issue	Description	Recommendation
Where and how are ventilation system emissions monitored?	When used to develop overall methane emission estimates, the optimal location for ventilation air monitors is at the point where ventilation air exhausts to the atmosphere.	If ventilation emissions are not monitored at the point of exhaust, emission data should be corrected based on estimated additional methane emissions between the point of measurement and the point of exhaust to the atmosphere.
Are ventilation system emissions monitored and/or reported for all mines?	In some countries, emissions are only reported for 'gassy mines'.	Estimates should be developed for non-gassy mines as well. Estimates can be prepared using information about the definitions of gassy and non-gassy mines and data on the total number of mines.
Are methane emissions from degasification systems reported?	Some countries collect and report methane emissions from ventilation and degasification systems, while others only report ventilation system emissions. Both emission sources must be included in emissions estimates.	If degasification system emissions are not included, those mines with degasification systems should be identified and estimates prepared on emissions from their degasification systems. Emissions estimates can be based on knowledge about the efficiency of the degasification system in use at the mine or the average efficiency of degasification in the country.

predicted level. This is increased to 33% higher or lower than the predicted level when used in countries other than the USA to take account of the range of mining operations and the differences in *in situ* methane contents worldwide.

Ideally it would be useful to be able to deduce methane emission levels from data based on coal rank and depth by deducing the *in situ* methane content (via Figure 2, say) and hence to the related methane emissions.[5] However, a recent UK study[6,7] has shown that large errors can be involved in even deducing *in situ* methane content. Clearly many other factors, particularly geological, come into play.

Emissions from Abandoned Deep Coal Mines

Some methane is also released from coal waste piles and abandoned mines. Coal waste piles are comprised of rock and small amounts of coal that are produced during mining along with marketable coal. There are currently no emission measurements for this source. Emissions are believed to be low, however, because much of the methane probably would be emitted in the mine and the waste rock would have a low gas content compared to that of the coal being mined. Emissions from abandoned mines may come from unsealed shafts and from vents installed to prevent the build-up of methane in mines. There is very little information on the number of abandoned mines, and data are currently unavailable on emissions from these mines. Most available evidence indicates that methane flow rates decay rapidly once deep mine coal production ceases.[7] In some abandoned mines, however, methane can continue to be released from surrounding strata for many years. In Belgium, France, and Germany, for example, several abandoned mines are currently being used as a source of methane which can be added to the gas supply system.[9] Due to the absence of measurement data for both coal waste piles and abandoned mines, no emissions estimates have been developed for these sources.

Methane flow rates decay rapidly once deep mine coal production ceases.[6] Underground measurements taken a few weeks after coal production has ceased are available together with the observation that old, sealed-off districts do not generally produce significant methane flows, other than during rapid barometer falls. When water pumping is halted on abandonment of a mine, water levels start to rise underground. Once the workings are flooded no further gas release or migration will occur. Abandoned mines therefore do not emit significant quantities of methane to the atmosphere and, at the most, these are estimated to be 1% of the normal ventilation emissions. This is almost insignificant compared with other emissions.

Emissions from Surface Mining

In surface mines, exposed coal-faces and surfaces, as well as areas of coal rubble created by mining operations, are believed to be the sources of methane. As in underground mines, however, emissions may come from the overburden, which is rubblized during the mining process, and underlying strata, which may be fractured and destressed due to removal of the overburden. Because surface mined coals are generally of lower rank and less deeply buried, they do not tend to contain as much methane as underground mined coals. Thus, emissions per tonne of coal mined are believed to be much lower for surface mines.

Indeed, little data exists on which to base emission from surface mines. The EPA have undertaken a study on a large Powder River Basin surface mine in Wyoming using long path FTIR spectroscopy. Other US studies have indicated emission contents in the range of 0.03 to $4 \, \text{m}^3 \, \text{t}^{-1}$ and an average value of $2.5 \, \text{m}^3 \, \text{t}^{-1}$ has been accepted. However, some recent figures obtained in the UK[6,7] suggest a lower value of $0.5 \pm 0.3 \, \text{m}^3 \, \text{t}^{-1}$.

As far as making accurate estimates is concerned, two possible approaches for estimating methane emissions from surface mining are suggested by the OECD Experts Group[11] similar to those developed for underground mining, but the results are more uncertain due to the absence of accurate emissions data.

(*i*) *Global Average Method.* As for underground mining, the simplest approach for surface mines—called the Global Average Method— is to multiply surface coal production by a range of emission factors representing global average emissions, as shown in the equation below:

$$\text{Low CH}_4 \text{ Emissions (tonnes)} = \text{Low CH}_4 \text{ Emission Factor } [\text{m}^3 \text{ CH}_4 \text{ (tonne of coal mined)}^{-1}] \times \text{Surface Coal Production (tonnes)} \times \text{Conversion Factor}$$

(2a)

$$\text{High CH}_4 \text{ Emissions (tonnes)} = \text{High CH}_4 \text{ Emission Factor } [\text{m}^3 \text{ CH}_4 \text{ (tonne of coal mined)}^{-1}] \times \text{Surface Coal Production (tonnes)} \times \text{Conversion Factor}$$

(2b)

Where the Low CH_4 Emission Factor equals $0.3 \, \text{m}^3 \, \text{tonne}^{-1}$ and the High CH_4 Emission Factor is $2.0 \, \text{m}^3 \, \text{tonne}^{-1}$. This does not include post-mining emissions.

Given the lack of information and measurements on methane emissions from surface mines, this range must be considered extremely uncertain, and it should be refined in the future as more data become available.

(*ii*) *Country or Basin Specific Method.* A second tier estimation of methane emissions—called the 'Country or Basin Specific Method'—can be used if additional information is available on *in situ* methane content and other characteristics of a country's surface mined coals. This approach enables a country to develop emission factors that better reflect specific conditions in their countries. Depending on the degree of detail desired, emissions can be estimated for specific coal basins or countries, using the equation below:

$$\text{CH}_4 \text{ Emissions (tonnes)} = \textit{In situ} \text{ Gas Content } [\text{m}^3 \text{ CH}_4 \text{ (tonne)}^{-1}] \times \text{Fraction of Gas Released During Mining (\%)} \times \text{Multiplier Reflecting the Contribution of Surrounding Strata} \times \text{Surface Coal Production (tonnes)} \times \text{Conversion Factor}$$

(3)

The *in situ* Gas Content represents the methane actually contained in the coal being mined, as determined by measuring the gas content of coal samples. Average values for a coal mine, coal basin, or country could be developed, depending on the level of detail in the estimate.

The Fraction of Gas Released During Mining represents the percentage of the *in situ* gas content that is assumed to be emitted during the mining process, as opposed to during post-mining activities. Estimates of this fraction will vary depending on a particular coal's characteristic methane desorption rate, local mining practices, and other factors. A likely range for this fraction appears to be between 60 and 100 per cent, based on recent studies.

The Multiplier Reflecting the Contribution of Surrounding Strata represents the possibility that more methane will be emitted during surface mining than is contained in the coal itself because of emissions from the strata above or below the coal seam. There is significant uncertainty about the potential contribution of methane from the surrounding strata in surface mines because of the likelihood that these strata have low gas contents and that much of the methane would have been released naturally prior to mining. However, one measurement study has found that the surrounding strata could increase emissions by as much as five times as compared with the gas contained in the coal. Selection of the multiplier should incorporate information about the coal characteristics, local mining practices (such as mine depth), and the geology of the basin. Based on current analyses, it appears that a reasonable range for this multiplier is between 1 and 5.

Post-mining Activities

Like surface mining emissions, there are currently few measurements of methane emissions from post-mining activities. In fact, many past studies have overlooked this emission source, while others have developed only rudimentary estimation methodologies. Two possible approaches for estimating emissions from post-mining activities are recommended by the OECD Experts Group.[11]

Tier 1: Global Average Method.   For the simplest estimates, a global average emission factor can be multiplied by coal production for underground and surface mining. It is important to distinguish between underground and surface mined coals because the gas contents are likely to be very different and hence emissions could vary significantly.

The recent OECD Experts Workshop[11] recommends emission factors of 0.9 to $4 \, m^3 \, tonne^{-1}$ for underground mined coal, based on recent studies and for surface mined coals emission factors of 0 to $0.2 \, m^3 \, tonne^{-1}$ are recommended.

Tier 2: Country or Basin Specific Method.   Emissions estimates can be refined if additional data are available on coal characteristics. This method may be preferable if higher tier methods have been used to estimate emissions from underground and surface mines.

The *In situ* Gas Content represents the methane actually contained in the coal being mined, as determined by measuring gas contents in coal samples. Average values for a coal mine, coal basin, or coal country could be developed, depending on the level of detail in the estimate.

Total Emissions from Coal Mining Activities

The total methane releases as a result of coal mining activities will be the summation of emissions from underground mining (ventilation and degasification systems), surface mining, and post-mining activities. To the extent that methane is recovered and used that would otherwise have been released to the atmosphere during coal mining, the recovered quantity should be subtracted from the emission total.

Data are readily available to develop general emissions estimates using the Tier 1 approach—the Global Average Methods for underground, surface, and post-mining activities. For these estimates, the only required data are country statistics on underground and surface coal production, which are available from domestic sources, such as energy ministries, or from the OECD/IEA, which publishes Coal Information and Coal Statistics.

On the basis of this type of analysis the global methane emissions has been estimated[12] to be 35 ± 10, with other estimates similar to this, IPCC 1992 used higher emission factors, now known to be erroneous and the value of $100 \, \text{Tg yr}^{-1}$ should be discarded. The current data suggests a value of about $45 \pm 10 \, \text{Tg yr}^{-1}$. Of this about 35% results from the coal industry in China, about 25% from that in the USA, and about 3% from the UK.

3   Technical Option for Emissions Control

All of the methods described above, with the possible exception of the Mine Specific Method, assume that all of the methane liberated by mining will be emitted to the atmosphere. In many countries, however, some of the methane recovered by mine degasification systems is used as fuel instead of being emitted. Wherever possible, the emission estimates should be corrected for the amount of methane that is used as fuel, by subtracting this amount from total estimated emissions.

In several countries, data on the disposition of methane recovered by degasification systems (*i.e.* whether it is used or emitted to the atmosphere) can be obtained from the coal industry or energy ministries. In Poland, for example, its mine degasification systems recovered 286 million m^3 of methane in 1989, of which 201 million m^3 was used and the remaining 85 million m^3 was emitted to the atmosphere.

Since the quantity of methane released is mainly a result of deep coal mining activities it seems likely that this activity is the best one to which to devote attention initially, as is indeed the case. Surface mining produces less methane over a greater distributed area and is much more difficult to handle technically. However, in view of the rapid expansion in opencast mining in many countries, much more attention must be devoted to this area. The approach taken would vary from country to country, but general principles have been noted and are given in a number of reports.

Techniques for removing methane from deep mine workings have been developed primarily for safety reasons, because it is highly explosive in air in concentrations between 5 and 15 mol%. These same techniques have been adapted in some places to recover methane so that the energy content of this fuel is not wasted. Methane emissions into the atmosphere can be reduced by up to 50–70% at gassy mines using available techniques. Emissions can potentially be reduced by up to 90%, depending upon the demonstration of additional technologies. However, generally this is not the case on a worldwide basis and overall recovery may be in the 0–20% region. Furthermore, about 10–15% of the methane present is removed in the mined coal, although a substantial part of this is burned while still in the coal.

[12] J. Lelieveld, P. J. Crutzen, and C. Brühl, *Chemosphere*, 1993, **26**, 739.

Important factors when considering options for reducing methane emissions from deep coal mining are: mine conditions (*e.g.* gassings); current mine gas systems; potential gas quality and use options; and technical and economic resources. In particular, the quality of gas that is recovered will determine the possible utilization options. Therefore, each of the four identified options is a coherent project based on recovering and utilizing a certain quality of gas.

Pre-mining Degasification. This strategy, often termed pre-draining, recovers methane from virgin seams before coal is mined. The advantage of this strategy is that methane is removed before the air from the mine workings can mix with it, and consequently a higher calorific value gas mixture is recovered (32–37 MJ m^{-3}). High quality gas will have a higher heating value and can also be used as chemical feedstock. Pre-mining degasification can be an in-mine or surface operation. When done inside the mine, boreholes can be drilled anywhere from six months to several years in advance of mining. The surface approach to pre-mining degasification requires more advanced technology and equipment, and therefore has higher capital costs than enhanced gob well recovery. These higher costs can be justified by the increased recovery of methane using surface drilling techniques.

Enhanced Gob Well Recovery or Post-drainage. The methods available are as follows: this strategy recovers methane from the gob area of a coal mine—the highly fractured area of coal and rock that is created by the caving of the mine roof after the coal is removed. Gob areas can release significant quantities of methane into the mine, and if this gas is recovered before entering the mine, ventilation requirements can be reduced. Typically, gob gas is diluted by mine air during production so a medium quality gas is obtained (11–29 MJ m^{-3}), which can be used in a variety of applications, including on-site power generation and residential and industrial heating. Enhanced gob well recovery can involve in-mine and/or surface wells, using existing, proven technology that is currently employed in many countries. The capital requirements are low compared with the potential for methane recovery.

Deep-mining Ventilation Air Utilization. Most mine gas is released to the atmosphere by the venting of in-mine air with large fans. For safety reasons, ventilation is necessary in deep coal mines. The recovery technology is basic, but operating costs can be high if the mine is gassy. The vented air is extremely dilute, at less than 1% methane but can be concentrated by a variety of methods and the use of membranes and pressure swing absorption techniques are promising. In addition, the use of such recovered methane as combustion air in turbines or boilers is feasible.

Application of Techniques. Methane emissions from coal mining constitute a wasted energy resource and often integrated recovery is used. Maximum utilization of this resource can involve a combination of two or three strategies, to provide an integrated methane recovery system. Technological and capital requirements are moderately high. However, effective use of these strategies can lead to the recovery of sufficient methane to justify these costs. Benefits of these strategies include improved mine safety and productivity and the recovery of a clean and convenient energy source.

The major challenge for the future, however, is to develop economically viable technologies for open-cast mining. Pre-drainage can be useful but additional techniques applicable to the mining operations are required.

Acknowledgements

We wish to thank Dr D. J. Buchanan and Dr D. Creedy (British Coal) for very helpful assistance. We are also particularly indebted to Ms Dina Kruger and Ms Kathleen Hogan of the US EPA and to other members of the OECD Experts Subgroup on Coal Mining. The section on Methodology draws heavily upon their work at their 1993 meeting.

Constructing Ecosystems and Determining their Connectivity to the Larger Ecological Landscape

J. CAIRNS, JR. AND R. B. ATKINSON

1 Introduction

Wetland construction is occurring throughout the world. Once the variety of ecological services provided by wetlands were elucidated,[1] many groups of people conceived the notion of constructing a wetland to perform a specific service in the hope of achieving a cost saving. Wetlands began to be constructed for treating sewage,[2] stabilizing shorelines,[3,4] controlling stormwater,[5] disposing of dredged material,[6] and treating acid mine drainage from coal mining operations.[7] Myriad case studies report a wide range of efficacy for these wetland construction efforts, but these projects benefit from the fact that the service to be provided by the construction is usually a stated goal of the project.

During much of the time that wetlands were being constructed for these specific functions, regulation of natural wetlands by the US Corps of Engineers (under Section 404 of the Clean Water Act of 1977) began to approve permits for wetland fills in exchange for constructed wetlands. Unlike most other 'single service' construction efforts, the goal here was replacement of the suite of services provided by the original wetland. By converting an upland site into a wetland that is structurally similar to the filled wetland, and preferably close to the filled wetland, all services would continue to be provided. Several recent reviews of

[1] J. H. Sather and R. D. Smith, 'Proceedings of the National Wetland Assessment Workshop', FWS/OBS-84-12, Fish and Wildlife Service, Washington, DC, 1984, p. 100.

[2] D. A. Hammer, 'Constructed Wetlands for Wastewater Treatment', Lewis Publishers, Chelsea, Michigan, 1989, p. 831.

[3] E. W. Garbisch, Jr., P. B. Woller, and R. J. McCallum, 'Salt Marsh Establishment and Development', US Army Corps of Engineers, Coastal Engineering Research Center, Ft. Belvior, Virginia, 1975, p. 110.

[4] E. D. Seneca, in 'Rehabilitation and Creation of Selected Coastal Habitats: Proceedings of a Workshop', FWS/OBS-80/27, ed. J. C. Lewis and E. W. Bunce, US Fish and Wildlife Service, Washington, DC, 1980, p. 58.

[5] L. W. Adams and L. E. Dove, 'Urban Wetlands for Stormwater Control and Wildlife Enhancement', National Institute for Urban Wildlife, Columbia, Maryland, 1984, p. 15.

[6] M. C. Landin and H. K. Smith, 'Beneficial Uses of Dredged Materials', Proceedings of the First Interagency Workshop, US Army Corps of Engineers Waterways Experiment Station, Vicksburg, Mississippi, 1987.

[7] R. P. Brooks, D. E. Samuel, and J. B. Hill, 'Proceedings of a Conference: Wetlands and Water Management on Mined Lands', Pennsylvania State University, University Park, Pennsylvania, 1985, p. 393.

these 'compensatory' constructed wetlands suggest that the constructed wetlands are generally quite inferior to the predisturbance condition of the natural wetland, although some services may be adequately replaced.[8,9] Other studies have found that wetlands can be constructed as an alternative ecosystem in degraded or damaged lands.[10] This discussion focuses on the rationale for constructing wetlands for multiple services in order to enhance restoration of surface mined areas.

2 Post-mining Site Reclamation

Current reclamation practices may yield sites with a limited ability to perform ecological services. Specifically, services associated with wildlife utilization, water quality enhancement, and hydrologic modification are not maximized. Many post-mining land uses include minimal vegetative diversity that is often limited to non-native members of two families, *Fabaceae* (legumes) and *Poaceae* (grasses). Wildlife utilization potential may be limited by minimal habitat diversity and cover, and a lack of standing water. Many populations may be fragmented by wide reclaimed strips, due to an inability or unwillingness to cross these areas.

Severe water quality problems may also limit site ability to perform ecological services. Low pH and elevated metal concentrations are often associated with surface mining. Wetland treatment systems may be appropriate for such areas since many researchers have found significant treatment efficacy afforded by current technologies.[7] For any given pollutant, a wetland may act as transformer, filter, or sink at any given time.[11] However, the ecological services of wetlands are not all necessarily compatible. The potential for food chain mobilization and biological magnification of surface mine toxicants in wetlands have not been adequately considered in the literature. The burden of proof regarding these ecological risks should fall to those proposing wetland construction for both toxicant removal and wildlife usage. Therefore, wetland construction sites in this study received water within US Environmental Protection Agency water quality standards.

Off-site Impacts

Restoration of a damaged ecosystem should not be limited to approximating local preimpact conditions, but should also consider mitigating damage within the landscape. This would include minimizing off-site degradation caused by the reclaimed ecosystem. Increased sedimentation and exaggerated water flow can accompany the large scale disturbance associated with surface mining in

---

[8] M. S. Race, *Environ. Manage.*, 1985, **9**, 71.

[9] J. A. Kusler and M. E. Kentula, 'Wetland Creation and Restoration: The Status of the Science', Island Press, Washington, DC, 1990, p. 595.

[10] J. Cairns, Jr., *Min. Environ.*, 1983, **5**, 32.

[11] C. J. Richardson, in 'Freshwater Wetlands and Wildlife', DOE Symposium Series No. 61, ed. R. R. Sharitz and J. W. Gibbons, Office of Scientific and Technical Information, Oak Ridge, Tennessee, 1989, p. 25.

montane regions. Sediment run-off into streams and increased suspended solids have been identified as 'the most destructive features' in surface mined areas with widespread surface disturbance and/or disturbance in steep terrain. The primary cause is removal of vegetative cover, and the sediment deposition in stream channels has deleterious effects on stream biota. Sediment loss from mined watersheds can exceed that for unmined watersheds by a factor of 1000. Sediment reduces light penetration and alters temperature in streams, reduces fish production as food organisms are buried and spawning grounds are filled, and can choke streams and increase potential for flooding.[12]

US Regulatory Agencies, Standards, and Practices

Current reclamation practices have their origin in the Surface Mining Control and Reclamation Act of 1977 (SMCRA; P.L. 95-87) administered by the Office of Surface Mining Reclamation and Enforcement (OSMRE) within the US Department of the Interior. In Virginia, the Division of Mined Land Reclamation (DMLR) has promulgated regulations and assumed primacy, and OSMRE retains a supervisory role. These extensive regulations were phased in over several years and exerted a dramatic influence on reclamation. Among its components, the SMCRA called for the replacement of the approximate original contour (AOC) following mineral removal, extensive monitoring of water quality parameters, minimal standards for revegetation, and the use of sediment control structures (usually referred to as sediment ponds).

One of the effects of the 1977 law, although perhaps unintended, was the rapid dewatering of reclamation sites. The return to approximate original contour helped drain much of each reclamation site. Saturated soils at the foot of a fill were also undesirable due to concern for AOC fill stability, since the structural integrity of fills could be weakened by significant wetting of the material. Another reason to channel water away from reclaimed sites was the occasional exposure of pyritic material, which led to generation of acidic and concentrated iron and manganese discharges from operations that exposed significant volumes of pyritic material. Any part of a reclamation site with a discharge point would necessitate a water quality permit (National Pollution Discharge Elimination System). Since most of the reclamation site was well drained, the species planted to revegetate were primarily obligate upland species, *i.e.* species that occur in wetlands less than 1% of the time.[13] Not surprisingly, these species failed to colonize the poorly drained portions of reclamation sites. With considerable pressure to maintain well drained and densely vegetated conditions at reclamation sites, reclamation designs included moderate to steeply sloped contours and rock drains to confine the water to a channel and quickly remove it from a site. Thus, sediment ponds were the only areas designed to retain moisture, but regulations allow sediment pond removal within two years of the start of reclamation.

[12] R. D. Hill and E. C. Grim, in 'Recovery and Restoration of Damaged Ecosystems', ed. J. Cairns, Jr., K. L. Dickson, and E. E. Herricks, University Press of Virginia, Charlottesville, Virginia, 1975, p. 290.

[13] P. B. Reed, Jr., 'National List of Plant Species That Occur in Wetlands: National Summary', US Fish and Wildlife Service, Washington, DC, 1988.

3   Constructed Wetlands in Coal Mining Regions of Southwest Virginia

Even though reclamation practices evolved to exclude saturated soils, language in Virginia DMLR regulations allows inclusion of 'depressions' for the benefit of water retention for wildlife. Section 480-03-19.816.102, backfilling and grading paragraph (h), merely states that depressions may be left for wildlife, sediment retention, and flow amelioration. Although ecologically sound, the lack of engineering guidance regarding design specifications proved prohibitive. Coal company reclamation personnel lacked a definition of a depression and how to construct this feature on a reclamation site. Requirements for an ongoing OSMRE-funded study conducted by our University Center for Environmental and Hazardous Materials Studies included: (1) providing coal companies with design specifications and (2) providing DMLR with compliance monitoring criteria.

As with most restoration research, practical application is the primary goal, though theoretical advances are often associated with well-planned studies. Practical considerations in this case included the willingness of coal operators to include constructed wetlands in post-mining reclamation plans. Several concerns had to be addressed. The constructed wetlands could come under US Army Corps of Engineers (Corps) regulatory authority, which could require notification of wetland fills if land use changes occurred in the future. In this region of Southwest Virginia, the primary reason these constructed wetlands might be filled would be remining. In that instance, a Corps permit to fill the wetlands might be issued so long as constructed wetlands were again a component of the post-remining reclamation plan.

Both DMLR and the coal companies expressed concerns over liability associated with accidental drowning by swimmers and fishers. This necessitated design modifications, and an agreement of a maximum depth of 1.2 m was reached. A second liability and DMLR regulatory considerations were associated with the construction of dams. The solution for this concern was to limit dam height to no more than 0.6 m and achieve the 1.2 m maximum depth primarily by excavating the depression.

A third concern of coal companies dealt with acid mine drainage (discussed earlier and at greater length in other articles in this volume). Some mining operations expose acid generating strata and other toxicants may also be present [iron and manganese are often associated with the sulfuric acid that is generated from contact with pyritic strata, and the solution is referred to as acid mine drainage (AMD)]. Current regulations do not allow bond release for sites where AMD treatment is occurring, and wetlands receiving AMD are considered treatment facilities.

Landscape Compatibility

A principal consideration in constructed wetland design is landscape compatibility. Cove hardwoods, hemlocks (*Tsuga canadensis*), and *Rhododendron* spp. communities that are associated with mountain streams in the unglaciated Allegheny Plateau become fragmented by strip mining. Restoration of this community requires: (1) halting environmental degradation and (2) establishing a community that can perform ecological services and lead, through succession, to an approximation of the predisturbance community.

After mining, the coves are replaced by broad, flat areas crossed by intermittent, often braided streams. Small depressions could be designed to retain water in this area. The historical presence of beaver (*Castor canadensis*) provided a natural model for a regionally appropriate wetland type. Dam construction by beavers in the region leads to small (less than 1 hectare), palustrine, open-water wetlands that follow a successionary pathway that generally includes submergent, emergent (herbaceous), scrub–shrub, and then forested wetlands. This sequence has often been abbreviated in the southern Appalachians as a result of sedimentation, first from deforestation around the turn of the century and then from surface contour (or opencast) mining.

The potential for wetland establishment on surface mined lands is provided by wetlands that formed 'accidentally' following surface mining before recent legislation. Prior to SMCRA, contour surface mining left many depressions. These areas collected sediment from the often poorly revegetated adjacent areas, filled with water, were colonized by wetland plants, and were a source of water for both upland and wetland wildlife.[14,15] In addition to satisfying both the hydrology and vegetation criteria of the legal definition of wetlands, the hydric soil criterion has also been met by virtue of demonstrating both low chroma (below two) and oxidized rhizospheres.[16,17] These wetlands are classified as palustrine, emergent wetlands.[18] Microbes, macrophytes, invertebrates, fish, amphibians, reptiles, birds, and mammals have been catalogued for many accidental wetlands. Waterfowl surveys from 1956 to 1985 listed 59 species for coal surface-mined wetlands in Illinois, USA.[19]

A Model for Wetland Construction

Review of natural and accidental wetlands in the region has provided a model for wetland construction, which is being field validated. Wetlands are being constructed in a series of three, each approximately 10×50 m and up to 1.2 m deep. The first (upstream) wetland in each series is expected to accumulate sediment quickly as revegetation progresses and to limit sedimentation in the next two wetlands in the series. Accidental wetlands were used as donors for

[14] A. A. Arata, *J. Wildl. Manage.*, 1959, **23**, 177.

[15] R. Bell, *Ill. Acad. Sci. Trans.*, 1956, **48**, 85.

[16] Environmental Laboratory, 'Corps of Engineers Wetlands Delineation Manual', Technical Report Y-87-1, US Army Engineer Waterways Experiment Station, Vicksburg, Mississippi, 1987.

[17] Federal Interagency Committee for Wetland Delineation, 'Federal Manual for Identifying and Delineating Jurisdictional Wetlands', Cooperative Technical Publication, US Army Corps of Engineers, US Environmental Protection Agency, US Fish and Wildlife Service, and USDA Soil Conservation Service, Washington, DC, 1989.

[18] L. M. Cowardin, V. Carter, F. C. Golet, and E. T. LaRoe, 'Classification of Wetlands and Deepwater Habitats of the United States', FWS/OBS-79/31, US Fish and Wildlife Service, Washington, DC, 1979, p. 103.

[19] J. R. Nawrot and W. D. Klimstra, in 'Animals in Primary Succession', ed. J. D. Majer, Cambridge University Press, Cambridge, England, 1989, p. 269.

hydric soils to provide both a seed source and a microbial community in order to hasten wetland establishment during the bond period. Soil amendments are only made to the second in each series since: (1) soil amendments to the first wetland probably would be buried by alluvium and (2) successful establishment in the second wetland should provide a source of propagules for the third wetland, thereby reducing construction costs and limiting disturbance to accidental wetlands.

Ecological Services

Wetlands are known to provide a diverse array of ecological services, many of which are closely linked with wetland functions. Included are wildlife habitat, water quality enhancement, and hydrologic modification. Maximizing performance of these three ecological services represents the goal for wetland construction on surface mined lands in our current study. Their potential relevance to the landscape is discussed below.

Wetlands provide a variety of benefits to wildlife. Many wetlands exhibit high primary productivity that supports wetland food webs and provides nesting sites for wetland fauna.[20] While these wildlife benefits derived from wetlands are recognized as being locally significant, wetlands constructed in natural drainageways on surface contour mined land may be positioned to perform additional services in a landscape and restoration context: (1) wetlands that connect unmined areas at elevations above and below the mined area may perform corridor services by providing an aquatic medium for some species and cover for others to pass through;[21] (2) wetlands may provide habitat amenities for wetland species and upland species in the region; (3) the attraction of both upland and wetland fauna to the constructed wetlands may lead to import of both upland and wetland propagules; (4) export of organic matter to aquatic communities downstream from constructed wetlands may be a more suitable substrate (and possess greater mass) than upland-derived allochthonous inputs.

The role of animals in restoration is understudied, but many researchers assign a great importance to animals in restoration.[22] One example is a study of turtles in freshwater wetlands.[23] Turtle productivity ranged from $7.3\,kg\,ha^{-1}\,yr^{-1}$ to $9.7\,kg\,ha^{-1}\,yr^{-1}$. Importance of turtles includes food for terrestrial species since roughly 40% of their productivity is egg production and since nest predation rates, *e.g.* by red fox (*Vulpes vulpes*) and raccoons (*Procyon lotor*) may exceed 95%. Turtles may recolonize disturbed sites because of their ability to move overland, which allows them to serve as vectors for wetland plants.[23] Majer[22] reviewed several faunal studies of surface mine restoration and concluded that certain plant species facilitate particular animal invasions, and those animals may alter sites to favor additional plant and animal immigrations. It seems likely, although difficult to quantify, that wetland corridors through a strip-mined area

[20] W. J. Mitsch and J. G. Gosselink, 'Wetlands', Van Nostrand Reinhold, New York, 1986, p. 539.
[21] R. T. T. Forman and M. Godron, *Bioscience*, 1981, **31**, 733.
[22] J. D. Majer, 'Animals in Primary Succession, The Role of Fauna in Reclaimed Land', Cambridge University Press, New York, 1989, p. 547.
[23] J. D. Cogdon and J. W. Gibbons, in 'Freshwater Wetlands and Wildlife', ed. R. R. Sharitz and J. W. Gibbons, US Department of Energy, Washington, DC, 1989, p. 583.

would increase floral and faunal establishment both in and adjacent to the wetlands.

The major pollutants from surface-mined coal sites are acid mine drainage and sediment. Although acid mine drainage is often not associated with surface mining, treatment of this pollutant is best performed by wetlands designed specifically for this function. Conversely, sediment removal processes may be compatible with wildlife services. Sedimentation rates should be highest during the earliest periods following reclamation and should slow as revegetation progresses in the watershed. The initially high sedimentation rates may improve constructed wetland development by providing a suitable substrate. Since soils in reclaimed sites are typically compacted by the heavy equipment, loose sediments in the constructed wetlands may facilitate aquatic macrophyte invasion and enhance vegetative cover and productivity.

Ecological services performed by constructed wetlands with regard to water quality may also exert positive effects at the landscape level. Since sediment ponds may be removed after two years following the start of reclamation, the wetlands may provide sediment retention services that would protect aquatic communities downstream. Standard reclamation measures may include nutrient applications, some of which could be transported through run-off and impact communities downstream. Wetlands have been shown to improve water quality by ameliorating excess nutrient inputs.[24]

Wetlands constructed on surface contour mined land have the potential to provide landscape level ecological services related to floodflow modification. The ability of wetlands to store flood water and to release that water slowly over time may provide two benefits. First, reduced flood peaks may limit disturbance of communities downstream. Second, slow release of water over time might help maintain hydric conditions by supplementing base flows.[25] The latter could facilitate faunal use further upstream, thus enhancing landscape connectivity.

Some data from accidental wetlands support the notion that constructed wetlands may perform these ecological services. To date, 94 plant species have been found associated with 14 accidental wetlands in our study. In June, the mean number of species per site was 17.77 ± 7.51, rising to 18.29 ± 7.49 in August. The presence of so many species that were not planted at sites of approximately fifteen years post-reclamation suggests that these areas may attract wildlife that bring more propagules, although causality has not been established. That is, we have no data to confirm whether most plant species were brought by animals or whether animals were attracted by the plant species diversity. Although somewhat circumstantial, evidence for corridor services is provided by the presence of several amphibian species within, and both upstream and downstream from, accidental wetlands. Upland and wetland, small and large mammal tracks were found in shoreline sediments. For example, evidence from

[24] R. P. Gambrell and W. H. Patrick, Jr., in 'Plant Life in Anaerobic Environments', ed. D. D. Hook and R. M. M. Crawford, Ann Arbor Science Publishers, Ann Arbor, Michigan, 1978, p. 375.

[25] G. G. Hollands, G. E. Hollis, and J. S. Larson, in 'Mitigating Freshwater Wetlands Alterations in the Glaciated Northeastern United States: An Assessment of the Science Base', ed. J. S. Larson and C. Neill, University of Massachusetts, Environmental Institute, Amherst, Massachusetts, 1986, p. 131.

tracks suggests that racoons exhibit nightly utilization of accidental wetlands as well as wetlands upstream and downstream.

Accidental wetlands appear to perform sediment accumulation and retention services. Preliminary results suggest that mean annual sediment accumulation rate is from 1 to 4 cm p.a., and greater during the first years after reclamation. No correlation has been detected between sedimentation rate and species richness in accidental wetlands; however, portions of accidental wetland sites with compacted soils, *e.g.* little or no accumulated sediment, exhibit low cover (preliminary biomass data also suggest lower biomass in portions of accidental wetlands having compacted soils). Accumulation rates may be lower in constructed wetlands due to better revegetation of adjacent upland areas in modern reclamation practices. However, recently reclaimed sites may comply with current revegetation standards and still exhibit significant soil loss, even after sediment ponds have been removed. Therefore, constructed wetlands may perform ecological services related to sediment trapping at both the local and landscape levels.

Evidence for hydrologic modification is perhaps less reliable. Standard deviation for mean water levels in accidental wetlands was low (3.4 cm), suggesting minimal water storage. However, accidental wetlands were located on fully revegetated mine sites with broad benches. Both factors could limit exposure of accidental wetlands to flood peaks and preclude opportunities to perform flood reduction services. Flood reduction services may be provided by constructed wetlands on newly reclaimed sites where these factors may be limited. The low standard deviation further suggests that accidental wetlands may exhibit fairly continuous discharge, which may help maintain flow in aquatic communities downstream.

Considerable temporal changes can be anticipated for both the constructed wetlands and the landscape in which they occur. Allochthonous materials (primarily sand, silt, and clay sediment deposits) are likely to comprise the bulk of the inputs through the first five years. Autochthonous inputs are likely to predominate during subsequent years as aquatic macrophyte cover and productivity increases. These abiotic and biotic processes have not been fully quantified for the accidental wetlands, and newer reclamation techniques limit the utility of accidental wetlands as a model for constructed wetland ontogeny; however, the likelihood that constructed wetlands will function as sinks for both inorganic and organic matter suggests that these systems will become shallower over time. Flood-intolerant species will slowly encroach around the margins and obligate wetland species (*e.g.* cattail, *Typha* spp.) may be out-competed by facultative species (*e.g.* wool grass, *Scirpus cyperinus*, and soft rush, *Juncus effusus*).[13] In addition to shifts in species composition, physiognamic changes in vegetation may lead to a change from emergent to scrub–shrub and/or forested wetlands.

Changes in the landscape can also be anticipated. Unpredictable human-induced disturbances, *e.g.* logging of adjacent areas, are quite possible in the region. Insufficient data on reclamation exist to predict successional direction for the strip-mined area. In addition, successional direction of constructed wetlands cannot be predicted from accidental wetlands with high accuracy due to the novel conditions present during constructed wetland development. As a result, the long-term need for ecological services, such as sediment trapping and hydrologic modification, cannot be anticipated. Ecological services related to habitat are likely to change as the landscape components change; however, the similarity to beaver pond ontogeny and position in the landscape suggest that habitat services should continue to be performed.

Acknowledgements

Support for this project was provided by the Office of Surface Mining Reclamation and Enforcement, US Department of the Interior.

The Discharge of Waters from Active and Abandoned Mines

R. J. PENTREATH

1 Introduction

Water from active mines is discharged in a controlled manner in the UK under a system of licences called consents. But waters from abandoned mines—which are discharged without controls—are a source of poor water quality in specific areas. The situation is a long-standing one in many cases, but mines are still being abandoned so that, generally, the situation continues to deteriorate. This trend needs to be reversed. The extent to which remedial measures can and should be taken with regard to long-abandoned mines nevertheless needs careful consideration and priorities need to be established.

Dealing with the problem is not helped by the legal position, which is briefly discussed. The nature of mine water is briefly described also, because it determines not only the effects which such waters have on the aquatic environment, but the difficulties which arise when trying to ameliorate such effects, which vary from the aesthetic to the toxic. The scale of the problem has yet to be fully evaluated in the UK, primarily because many of the effects are most acute in streams and upper reaches which are not routinely monitored, and are thus unclassified. Nevertheless, some attempt has been made to characterize the scale and nature of waters affected by abandoned coal and metal mines and suggestions made as to how the situation could be improved.

2 The Nature of Mine Water and its Potential Environmental Impact

Deposits of coal, and of minerals, lie at various depths and have been exploited by man for a very long time. Even without human intervention, however, both surface and groundwaters have permeated such areas, leaching metals and other substances in the process. But mining increases the surface area of the deposits, facilitates and increases the passage of groundwater, and may thus accelerate the rate, and change the scale, of these leaching processes.

Water enters most mines in the UK. In working mines it has to be removed by pumping. In addition to the direct downward movement of rainwater reaching the underlying aquifers, water may also enter via faults, gallerys, and adits, many of which may extend well beyond the surface watershed of the catchment. The quantity of water can thus be very large and very variable. The chemical nature of such waters varies from mine to mine, but a common feature is that of a reddish-brown suspension due to the presence of iron minerals. Another common feature of such mines is the presence of iron pyrites which, upon prolonged contact with water, dissolves to form sulfuric acid. This can lead to the leaching of other metals which are naturally present. The final waters emerging from a mine may therefore be acidic, laden with metals such as cadmium, copper, and zinc, plus suspended materials which co-precipitate out as a highly coloured floc. Another not uncommon feature in some mines is that of 'saline' water—the salts being of chloride or sulfate.

Treatment facilities are installed and operated at working mines to reduce the potential impact of such waters—together with effluent arising from pithead activities. Not all mine water is necessarily of poor quality, however; indeed some of it is very good and is used to offset the effects of poor quality surface waters by providing additional dilution. Some mine water may also be used, after treatment, for potable supply.

The Chemistry of Mine Waters

The nature of mine waters, like those of surface waters and other groundwaters, varies very considerably. Different areas exist where waters draining from mines are alkaline, moderately or highly saline, alkaline and ferruginous, or acidic and ferruginous. The nature and effect of such waters can also differ, within the same mining complex, between that arising from shallow level workings and adits and that pumped or emerging from deeper levels. Understanding the nature of such waters is an essential preliminary to dealing with the discharges of individual mines. This in turn is usually related to the hydraulic features of the mine. And whilst it is difficult to generalize on this subject, it is useful to examine the causes of one of the most common features—that of ferruginous waters—emanating from mines, particularly coal mines.

Ferruginous mine waters are caused by the oxidation of iron pyrites (pyrite), which is a mineral form of iron sulfide; superficially it resembles gold in appearance, hence the name 'fools gold'. Iron pyrites is common in both the coal itself and in the mudstones, of marine origin, which overlie the coal seams. It is also common in metal mines. Up to 10% of coal layers may consist of this mineral. They may be continuously or sporadically exposed to air in near-surface levels, but in deeper workings such strata will have been below the water table. When the water table is lowered by pumping, these strata become exposed to air. The iron pyrites then rapidly oxidizes, although such oxidation can take place in a variety of ways and via a number of intermediate chemical products, depending on the precise environmental conditions. Factors which are known to influence the rate and extent of the oxidation reactions include the sulfide mineral content, its morphology, the availability of oxygen, and the ferric ion concentration. Several of the rates of oxidation are also greatly increased by the catalytic activity of bacteria. Other sulfide minerals present may undergo a similar series of chemical reactions when exposed to air and water. The oxidation creates acidic conditions, with the result that sulfuric acid is produced in various quantities and at different rates. This acid may then cause other minerals to dissolve.

In underground workings the pumping of mine water reduces the rate at which leaching occurs from exposed surfaces. Acidic mine waters are treated to neutralize them—if only to protect mining machinery! When mining operations cease, however, and the pumping stops, the water table rebounds to its natural level—or to a new level as a result of the mining operations. This flooding of the exposed seams stops the oxidation of the iron pyrites, but brings into solution the sulfuric acid and the iron sulfates which are the products of the oxidation reactions. The result of this depends on the nature of the rock strata. If they are calcareous, and particularly of limestone, the mine water may be neutralized; such waters usually have a reduced iron content. If they are not calcareous, however, the mine water may become highly acidic—as low as pH 1 or 2—and become even more loaded with iron, and often with manganese.

When the water finally reaches the surface it may emerge via old adits, emerge as springs, or as seepage through the ground or even through the bed of an existing river or stream. As it emerges it may well be clear and almost colourless, because the underground water is low in oxygen and the iron in solution. As this water mixes with the air—which may occur before it emerges above ground—and with oxygenated water, the iron rapidly oxidizes from the ferrous to the ferric form and precipitates out as an orange deposit. In shallow mines, or in adits set in higher ground, such cycles may be repeated continually as the groundwaters fluctuate. In deeper mines connections may be made with underground aquifers. Quite frequently the history and extent of mining is such that neither the hydraulic conditions nor the chemical state of the water can be easily or accurately predicted once the last mining activity ceases.

Similar chemical reactions also occur in colliery and metal mine spoil tips above the ground, such that run-off from them may be acidic and ferruginous. A further problem, however, is that they are a source of particulate material, usually of very fine, often colloidal, clay and shale particles which in turn may carry other chemicals, particularly metals, with them. Thus factors such as rainfall can affect the natural variation of waters in surface adits, and the rates of leaching within surface spoil tips. Discharges from active mines are controlled in order to prevent receiving waters from such effects.

Biological Impacts

The impacts on aquatic communities of untreated mine water may not be immediately apparent, but can be of serious environmental consequence. The observable biological effects include: (1) depletion of numbers of sensitive, and diversity of all, free swimming and benthic (bottom dwelling) aquatic organisms; (2) loss of spawning gravel for fish; and (3) direct fish mortalities, particularly of natural game (salmon and trout) fish. A range of less readily observed sub-lethal effects may also occur.

Clear streams can turn into highly ochreous ones of a vivid orange appearance. Such discharges make rivers virtually fishless by coating the river bed with precipitating iron hydroxides. Depletion of the numbers and diversity of benthic (bottom dwelling) species occurs because the precipitate has a smothering effect, reducing oxygen, and covering the river bed with iron oxides. This process also reduces the extent of spawning gravels for fish breeding, by occluding the interstices of the gravels with fine sediment, and therefore limiting the availability of nursery streams. Natural game fish populations are particularly susceptible to such pollution. The low pH can be directly toxic, causing damage to fish gills. It can also solubilize metals, not only those which emerge from the mine water, but those—such as aluminium—which become dissolved within streams, and which are also toxic to fish.

Perhaps the greatest impact of mine water pollution occurs in the smaller streams which are not usually classified under the river quality assessment schemes. These streams, which typically form the headwaters of rivers, are vitally important as fish breeding grounds and nursery areas for developing juveniles.

Impacts on Other Water Users

Other impacts of untreated mine waters include the imposition of restrictions on legitimate users of the water body, who may find the water unsuitable for irrigation, livestock watering, industrial, or potable water supply. There may also be significant consequences for shellfisheries, conservation areas, and for recreation and tourism.

The aesthetic impacts which ferruginous mine waters cause to rivers and streams, by the presence of a high colouration, immediately reduces the amenity value of an area. A direct consequence of this visual damage is a reduction in the use of a waterbody for recreational and watersport activities. Again, this reduces the value of the water resource to the local community. An impairment of the quality of a river because of mine-water pollution may also render it unsuitable for industrial and potable water supply, and often unsuitable for irrigation.

Predicting the effect on water quality as a result of mine closure is extremely difficult: the time taken for groundwater to rebound to a more or less equilibrium value can take from a few months to several years. Part of this is due to the point of entry, particularly via various mine shafts. Indeed a knowledge of and provision for looking after mine shafts is an integral part of dealing with the problem. Poor planning decisions are often made because of mine shaft location, and they are often inadequately infilled prior to redevelopment of the land. Shafts are often left uncapped, which not only leads to water entry, but to the use of the shafts for fly tipping an unknown quantity of potentially polluting chemicals and materials.

Finally, it is also important to consider that changed flow regimes which are a consequence of the cessation of pumping can affect flow rates of rivers on the surface. Changed flow patterns can then affect the availability of water for abstraction, or can lead to localized flooding problems as old wells and springs become reactivated, or exacerbate existing flooding problems within vulnerable downstream areas. Changes in flow patterns and groundwater levels may exceptionally render natural slopes unstable, resulting in landslides. Subsidence may also occur due to the softening of earth and mudstone. Methane gas may be forced to the surface. Flooding may occur.

As with all other situations involving the pollution of a natural resource, there is an economic cost as well as an environmental one. In this case, there are many economic costs. The reduction in the quality of water will affect the wide variety of uses which are demanded of it and ultimately there is a price to pay. The environmental consequences thus cannot be separated from the economic impacts and the two must be viewed in parallel.

3   Active and Abandoned Mines

It is evident that mine water is potentially very damaging. Water emanating from working mines is therefore carefully controlled, the (usually) acid waters neutralized, suspended solids removed, and the treated effluent then discharged under licence. Discharges from working mines in England and Wales are controlled through consents issued by the NRA under the Water Act 1989, since consolidated into the Water Resources Act 1991. Similar provisions exist elsewhere in the UK. Such consents generally include conditions relating to the quality, quantity, discharge regime, and monitoring of the mine water. Separate consents are given for discharges from related above-ground activities. Once the mine is no longer operational, the mine owner can ask for his consents to be revoked. Should pollution subsequently occur, there is then limited control over the discharges, for two reasons. Firstly, with regard to committing a pollution offence, the principal offence under Section 85(1) of the Water Resources Act 1991[1] (previously Section 107(1) of the 1989 Water Act), is that a person:

'. . . *causes or knowingly permits* any poisonous, noxious or polluting matter or any solid waste matter to enter any controlled waters'.

A defence under Section 89(3) of the Water Resources Act 1991 is that:

'A person shall not be guilty under Section 85 by reason only of his *permitting* water from an abandoned mine to enter into controlled waters'.

As can be seen, the defence relates only to *permitting*, which in the case of long-abandoned mines implies that action need not be taken to ameliorate the effect of past practices. An *abandoned mine* is not defined in the 1991 Water Resources Act, nor in any other relevant legislation. However, a *mine* is said to have the same meaning as that in the Mines and Quarries Act 1954.[2] It could be argued, perhaps, that the act of abandoning a mine *causes* pollution if, subsequent to such action, contaminated mine water enters controlled waters. Nevertheless, it is likely to be only one link in a chain of events. This was essentially the basis for the only successful prosecution in relation to mine closure—that of Lockhart *versus* National Coal Board (NCB) in Scotland in 1981. The case involved pollution of surface water resulting from the closure of a mine which had operated from 1951 to 1977. Under Section 22(1) of the Rivers (Prevention of Pollution) (Scotland) Act 1951,[3] the Crown argued that the NCB had *caused* pollution and that they could prove that they had '. . . carried out

---

[1] Water Resources Act 1991, HMSO.
[2] Mines and Quarries Act 1954, HMSO.
[3] Rivers (Prevention of Pollution) (Scotland) Act 1951, HMSO.

some active operations or chain of operations the natural consequence of which . . .' was to cause polluting matter to enter a stream: essentially, that the pollution resulted from a hole being dug which, after mining operations had ceased, filled with water and caused pollution. This was described by the court as '. . . the result of the respondent creating the latent danger and then leaving it to the mercy of the forces of nature, which activated the danger'.[4] Although helpful, this case does, however, illustrate the legal difficulties which arise when mines with a very long history are abandoned, particularly where they have been inter-connected with previously dug workings, many of which will be at a shallow level and may therefore be the principal contributors to the final poor quality water.

In the case of the Wheal Jane tin mine in Cornwall in 1992, because of the complicated history of the mine, the NRA—after taking detailed legal advice—concluded that it would not be successful in bringing a case against Carnon Consolidated Ltd., the mine's owners, for causing pollution in abandoning the mine. Thus the only purpose in attempting to bring about a prosecution would have been to demonstrate deficiencies in the law and, because of its capacity as a 'Crown prosecutor', this would not have been a correct use of the NRA's position, and it has therefore sought to secure legal changes by making direct approaches to the Government. (The Wheal Jane incident is discussed further in the next article). In their document 'This Common Inheritance The Second Year report',[5] the Government stated that it was considering '. . . the framework of legal responsibility for pollution from abandoned mines'.

In the case of coal mine closures—for mines operated by British Coal—when the closure of a colliery is announced, British Coal undertakes a structured analysis of the issues relating to the closure which stems primarily from safety considerations in relation to the impact on adjacent working mines. But it also addresses environmental and land ownership issues. The cessation of pumping is an important element of the closure programme and the matter is discussed with the NRA.

Where a colliery closure programme is implemented, 'due regard' is given by British Coal to the future activities and the need for a consented discharge to controlled waters. A proportion of mines and associated tipping lands are held on a leasehold basis and the requirements of the lease may restrict the opportunities available. In freehold situations British Coal is generally required by specific planning consent conditions, or by an arrangement with the Department of the Environment with respect to collieries closed since 1 April 1990, to restore colliery surfaces after abandonment unless alternative forms of development are accepted by the Planning Authority. When mines are not closed permanently they are retained on a 'care and maintenance' basis and therefore pumped and ventilated, and subjected to regular inspection.

The second control available relates to possible remedial action. Where pollution is occurring the NRA has general powers, under what is now Section 161 of the Water Resources Act 1991, to carry out works which may be necessary to prevent the contaminated water from entering controlled waters, or to remove, remedy, or mitigate any pollution, and restore the waters to their previous state.

[4] 'Water pollution from abandoned mines', Water Law, July 1992, pp. 119–120.
[5] This Common Inheritance: The Second Year Report, HMSO, October 1992.

The NRA is also entitled to recover expenses reasonably incurred in such work from those who caused or knowingly permitted the pollution *but* not

'from a person for any works or operations in respect of water from abandoned mines which that person *permitted* . . . to enter controlled waters'.

Thus, taking this and the situation with regard to prosecution together, the NRA and similar regulatory bodies are somewhat limited with regard to what they can do. Prosecution for polluting is difficult, and if it fails, the cost of remedial action would have to be borne by the regulatory authority. As the abandonment of a mine is often concomitant with its owners being in financial difficulties, any reimbursement of costs—even if prosecution was successful—is likely to be limited anyway. But an owner may have other sources of income, or have the capability of actual or potential development of the site from which an on-going revenue may be raised. The mine could also be re-opened within a relatively short period of time. The powers available to the NRA under Section 161 are nevertheless clearly of value in those circumstances where a polluting event has occurred and action can be taken to prevent or remedy the situation without any future commitment falling on either the NRA or the owner, providing that the funding is available. More difficult are those situations where remedial action would require long-term management of works installed to prevent further pollution: the NRA could rapidly acquire a long list of sites to manage in this way.

There is also the question of cleaning up the site above the ground, and the possibility of a mine being mothballed for subsequent re-opening. Such matters involve complicated issues with regard to land ownership, mineral and development rights, mining rights, and the planning law. Briefly the situation is as follows.

Ownership, Mineral Rights, and Mining

Mineral rights for precious metals and for uranium in the UK are all owned by the Crown, but with the exception of small gold mines in Wales, no mines have been worked exclusively for these metals in modern times. For coal, mineral rights rest with British Coal at present. For most other minerals ownership of land may also carry with it the ownership of the mineral rights, but in many cases the two have become separated and the mineral rights are owned separately.

A further complication is that neither the owner of the land nor the holder of the mineral rights may necessarily be the developer of a mine. In order to develop a mine in such a situation it is necessary for the would-be miner to apply to the owner for the lease of the land, the possessor of the mineral rights—to whom he will pay royalties—in order to exploit them, and to the relevant Planning Authority in order to develop a mine on site. In Cornwall the situation has been particularly complicated by Stannary Parliaments and Stannary Laws.

Planning and Remediation

Where a Mineral Planning Authority decides, after the necessary consultations, to grant planning permission for an application to work minerals—including underground mining—such permissions are normally subject to a number of attached conditions. These are likely to include requirements to minimize or prevent environmental effects (including visual impacts or disturbance to surrounding areas) during the operation of a site and to ensure reclamation of the site to a beneficial use when working has ceased. Mineral permissions are now time limited, so that if a mine has been abandoned for the purposes of the Mines and Quarries Act 1954, then it would depend on the time of the planning permission as to whether anyone wishing to re-open the mine would have to apply for a new permission to do so. In some circumstances the applicant/operator may also have entered into a voluntary agreement (now termed a 'planning obligation') with the mineral planning authority under Section 106 of the Town and Country Planning Act 1990.[6]

International Commitments

As a member of the European Community (EC) the UK Government is required to comply with EC legislation. Because of the metalliferous content of many mine waters, the NRA in England and Wales has specific responsibilities through the Water Resources Act 1991 with regard to their effect on compliance with EC Directives relating to dangerous substances in surface and underground waters. Of specific interest are the two metals cadmium and mercury. The relevant Directives have been implemented into national law by means of the Surface Waters (Dangerous Substances) (Classification) Regulations 1989 (SI 2286)[7] which specifies standards for freshwater (DS1) and coastal waters (DS2) within respective classification schemes. The NRA has been directed by the Department of the Environment (DoE) to perform various duties with regard to SI 2286, primarily relating to the consenting of discharges containing the substances listed, and the implementation of a suitable monitoring and analysis programme by which compliance with the classified objectives can be demonstrated.

When the NRA has reason to believe that surface waters are liable to fail the requirements of an annual mean standard, it has to provide the DoE with all relevant information as to the nature and circumstances of the reasons for failure, and the steps that the NRA has taken, or proposes to take, to restore the quality of the water. If such steps are unlikely to be effective within 12 months, the NRA has to provide the Department with such information as will allow the Secretary of State to '. . . determine in relation to any relevant discharge to those waters an appropriate emission standard in accordance with the relevant Council Directive'. It has been the practice, for the previous Water Authorities—and thus subsequently the NRA— to include in their monitoring programme those sites influenced by discharges from mining operations, but not necessarily those close to contaminated land. The NRA has also to send to the DoE by 30 April each year information on the discharges, sampling, and analysis for the previous calendar year, and details of variations and additions to related discharge consents.

---

[6] The Town and Country Planning Act 1990, HMSO.

[7] The Surface Waters (Dangerous Substances) (Classification) Regulations 1989, Statutory Instrument No. 2286, HMSO, 1989.

Other EC Directives of relevance to the topic of mine waters are the Freshwater Fisheries Directive,[8] the Abstraction of Drinking Water Directive,[9] the Shellfish Water Directive,[10] and the Groundwater Directive.[11]

4 The Scale of the Problem from Abandoned Mines

It might be thought that, with legal requirements in place to record abandoned mines, estimating the number of such mines should be a straightforward exercise; but this is not the case. The Mines and Quarries Inspectorate holds only non-coal records, and the data are not in a form which is easily accessible to the public. There are no records of any mines prior to 1872. British Coal holds its own data base of approximately 10 000 abandoned mine workings, but this figure is only an estimate and complications arise as a result of recent trends to re-open mines for commercial, recreational, and educational purposes.

The number of discharges from coal mines causing significant pollution, *i.e.* subjects of complaint, deterioration in water quality, and failure of non-statutory River Quality Objectives (RQOs) in England and Wales is about 100, affecting about 200 km of rivers. This does not include less serious discharges, of which there are many, or natural ochreous discharges.

As with coal, the mining of metals has a very long history in some areas, and many mines have been more or less continually worked for centuries. As a result, underground workings can be extremely complex and the extent of them is rarely fully known: above ground contamination with various metals can be equally extensive, and again this is rarely fully known. Entire water catchments may be affected, and delineating the precise source of contamination is in many cases virtually impossible, other than to conclude that the contaminant is ubiquitous within a very large area.

Certain regions therefore have problems with waters emanating from both abandoned metal and coal mines. Some 400 km of classified rivers are affected in England and Wales. Both the nature and extent of the problem varies, and its full scale is not easy to determine. In Devon and Cornwall, which have the principal problems with water from abandoned metal mines, there are at least 1700 abandoned mine-workings, many of which were small operations producing ore for a short time only. At present there is only one metal mine still in operation, on the Red River in Cornwall, and this is currently not causing significant pollution problems. Most mines produced several minerals—principally copper, tin, zinc, arsenic, lead, and some silver—with many minor minerals also being present. Some 22 catchments in the area are affected by non-ferrous mining activities, with 212 km of river being significantly affected. And this does not include the lengths of unclassified, unmonitored, streams—many of which are significantly affected.

---

[8] 'The Quality of freshwaters needing protection or improvement in order to support fish life', (78/659/EEC), 1978.

[9] 'The quality required of surface water intended for the abstraction of drinking water', (75/440/EEC), 1975.

[10] 'The Quality required of Shellfish Waters', (79/923/EEC), 1979.

[11] 'Protection of Groundwater against pollution caused by certain Dangerous Substances', (80/68/EEC), 1980.

Some discharges are semi-saline, some are warm, and some contain substantial quantities of naturally occurring radionuclides.

Wales, too, has had serious problems with abandoned metal mines—problems equally as serious as those arising from abandoned coal mines. There are over 500 abandoned metal mines in west and north Wales, the majority arising from the mining of lead, zinc, and silver in North Ceredigion near Aberystwyth and in north east Wales. In Meirionnydd, copper and gold were extracted, the gold being associated with mineralized veins containing zinc. In all cases the most serious pollution problems are related to the discharge of contaminated water draining from the abandoned mine workings. The most affected major rivers are Ystwyth and Rheidol near Aberystwyth. The poor quality is due to zinc pollution derived from underground drainage water from the abandoned mines. There are also significant problems with water quality on the Conwy, upper Teifi, upper Towy, upper Dovey, and upper Mawddach. At Parys Mountain, the Afon Goch is seriously polluted with elevated copper concentrations.

Since the NRA was set up, much effort has been expended to quantify further this inherited problem, with surveys being carried out particularly in the south west corner of England, and in Wales. Such work is being integrated with other studies on the not-unrelated problem of contaminated land, and will be evaluated through a process of catchment management planning. This therefore naturally raises the questions of what can be done, when, and how the residual water quality problems can be tackled.

5   Some Thoughts for the Future

There are several positive actions which could usefully be taken. These include a clarification of the law with respect to what constitutes an abandoned mine, the need to inform the relevant environmental agencies in good time of the intention to abandon a mine, and full allowance in the planning system for the future opening and closing of mines, including their re-opening. With regard to the standing of long-abandoned mines, however, it is clearly impractical to attempt to ameliorate their effects, by whatever means, until their relative contribution to poor water quality in different catchments has been fully assessed. It is suggested, nevertheless, that priority needs to be given to those mines which: (1) are a cause of breaching a surface water quality standard; (2) can be shown to be a significant contributor to the annual input of toxic chemicals into coastal waters; or (3) are a unique cause of poor water quality in an otherwise good quality river.

There is also the question of what, practically, can be done. It is clear that adequate provision needs to be made for the consequences of mine closure as soon as mines are opened, and some form of arrangement is needed for the long-term running of treatment plants, where necessary, when a mine has been closed in order to safeguard receiving water quality. There are several options available to deal with the polluting aspects of mine waters. Much of the ameliorative research and experimental work has been carried out in the USA; less is known about similar work in Europe. But, the NRA itself is extensively involved in research and development for the treatment of abandoned mine water.

The principal objectives of treating abandoned mine waters are to remove the iron floc and associated metals, and to adjust the pH. The potential treatments fall into three categories, as follows:

(1) *Physical systems*—which are processes in which oxidation of the water is accomplished through engineered cascades, together with facilities for sludge settlement. The costs involved are largely capital expenditure, but revenue costs arise for the disposal of the contents of settlement tanks and desludging processes.

(2) *Chemical systems*—which are of two sorts: active processes, which are expensive to run because of both the cost of the treatment chemicals, and the disposal of the resulting sludge; and passive processes, which are used in the USA and appear to be much less expensive.

(3) *Biological systems*—which are processes that include bacterial oxidation and the use of reed beds. Recent work in the USA has demonstrated that the use of reed beds is a relatively low cost approach.

The biggest problem, however, is how such work should be paid for, and by whom? Within the UK this still requires a wider national debate.

Environmental Best-practice in Metals Production

A. WARHURST

1 Introduction

Public policy to promote technical change and foster economic efficiency, rather than environmental regulation alone, is more likely to achieve sustained and competitive improvement in the long-term environmental management of our non-renewable natural resources. Three key issues are explored in this article.

The Relationship between Production Efficiency and Environmental Performance

There is growing evidence that technical change, stimulated by the 'Environmental Imperative', is reducing both production and environmental costs to the advantage of those dynamic companies that have the competence and resources to innovate. Such companies include mining enterprises in developing countries, as well as trans-national firms; but the evidence is strongest for large new investment projects and greenfield sites. In older ongoing operations, environmental performance correlates closely with production efficiency, and environmental degradation is greatest in operations working with obsolete technology, limited capital, and poor human resource management. The development of the technological and managerial capabilities to effect technical change in those organizations would lead to improved efficiencies in the use of energy and chemical reagents and in waste disposal, to higher metal recovery levels, and better workplace health and safety. This in turn would result in improved overall environmental management.

The Economic and Environmental Limitations of Regulation

Currently, the environmental performance of a mining enterprise is more closely related to its capacity to innovate than the regulatory regime within which it operates. Although international standards and stricter environmental regulation may not pose problems for the economics of new mineral projects, there could be major costs and challenges involved for older, and particularly inefficient, ongoing operations. Controlling pollution problems in many of these cases requires costly add-on solutions: water treatment plants, strengthening and

133

rebuilding tailings dams, scrubbers, and dust precipitators, *etc*. Furthermore, in the absence of technological and managerial capabilities, there is no guarantee that such items of pollution control—environmental hardware—will be incorporated or operated effectively in the production process. In some instances such requirements are leading to shut-downs, delays, and cancellations as well as reduced competitiveness. When mines and facilities are shut-down the clean-up costs frequently get transferred to the public sector, which, particularly in developing countries, has neither the resources nor technical capacity to deal with the problem effectively. In most countries, perhaps with the exception of the USA, the lack of retrospective regulation means the pollutee-suffers-and-pays principle is alive and well. This is not in itself, however, an argument either against regulation *per se* or for the global diffusion of Superfund legislation. (The Superfund legislation enables blame for environmental damage to be apportioned to a selected one of many past mine owners and for that company to be charged an estimated cost for work which government contracts-in to clean up and rehabilitate the damaged site.)

The Case for an Environmental Management Policy

The implication of this analysis is that to ensure competitive and sustainable environmental management practices in metals production, governments need to embrace public policy which goes beyond traditional, incremental, and punitive environmental regulation. The latter, in the old 'environmental protectionist' mode, tends to treat the symptoms of environmental mismanagement (*i.e.* pollution), not the causes (*i.e.* lack of capital, skills, and technology and the absence of the capability to innovate). The challenge will be for governments to ensure that companies operating within their national boundaries remain sufficiently dynamic to be able to afford to clean up when operations cease and to innovate to improve economic efficiency and environmental management in the meantime. Governments need policy tools which enable them to predict 'corporate environmental trajectories' and pick up the warning signs of declining competitiveness and impending mine close-down to ensure sufficient resources are available for the environmental management of mine 'decommissioning'. Policy mechanisms need to be developed to promote technical change and to build up the technological and management capabilities to innovate and manage the acquisition and absorption of clean technology. The privatization of the state sector and the liberalization of investment regimes in many developing countries, such as Angola, Mozambique, Namibia, Botswana, Bolivia, Peru, and Chile, with their emerging emphasis on joint-ventures and inter-firm collaborative arrangements, provides new opportunities for the diffusion of both competitive and environmentally sound best-practice in metals production.

Public policy to promote technical change and, complementary to that, to improve economic efficiency, respects the interplay between the environmental and economic factors that constitute a sustainable development approach to the long-term environmental management of our non-renewable natural resources. Environmental regulation at best provides only one element of a public policy for environmental management.

2 The Limitations of Environmental Regulation and Challenge for Public Policy

Regulatory frameworks for safeguarding the quality and availability of land, water, and air degraded as a result of mining and mineral processing activities are growing in number and complexity. This has particularly been the case in the major mineral producing countries of North America and Australia, as well as Japan and Europe. The norm in environmental regulation is that governments set maximum permissible discharge levels or minimum levels of acceptable environmental quality. Such 'command and control' mechanisms include: Best Available Technology standards, clean water and air acts, Superfunds for clean-up and liability determination, and a range of site-specific permitting procedures which tend to be the responsibility of local government within nationally approved regulatory regimes. Such 'command and control' mechanisms tend to rely on administrative agencies and judicial systems for enforcement. Three issues are relevant regarding the appropriateness of industrialized country's environmental regulations to reduce environmental degradation and improve environmental management practices in metals production.

First, there is a trend away from a 'pollutee suffers' to 'polluter pays' principle. However, it remains the case that the polluter pays only if discovered and prosecuted, which requires technical skills and a sophisticated judicial system, and that may occur only after the pollution problem has become apparent and has caused potentially irreversible damage. This highlights the tendency of such environmental regulations to deal with the symptoms of environmental mismanagement (pollution) rather than its causes (economic constraints, technical constraints, lack of access to technology or information about better environmental management practices). This can be serious in some instances because once certain types of pollution have been identified, such as acid mine drainage, it is extremely costly and sometimes technically impossible to trace the cause, rectify the problem, and prevent its recurrence. Certain environmental controls may only work if incorporated into a project from the outset (*e.g.* buffer zones to protect against leaks under multi-tonnage leach pads and tailings ponds).

Second, Best Available Technology (BAT) standards may be appropriate at plant start-up, but their specified effluent and emission levels are not necessarily achievable throughout the life of the plant, because technical problems may arise and there may be variations in the quality of concentrate or smelter feed, *etc.*, if supply sources are changed. Moreover, there are serious implications for monitoring. It would be also erroneous for a regulatory authority to assume that standards are being met if a pre-selected item of technology has been installed. Ongoing management and the environmental practices at the plant are likely to be important determinants of 'best environmental practices'.

Third, related to points one and two above, BAT standards and environmental regulations of the command and control type tend to presume a static technology—a best technology at any one time. This tends to promote incremental add-on controls to respond to evolving regulation rather than to stimulate innovation. This acts as a disincentive to innovate by equipment suppliers, the mining companies and metal producers. Their innovation, which has required substantial research and development resources, may be superseded by some regulatory authority's decision about what constitutes BAT for their particular activity. BAT gives the impression of technology being imposed from outside the firm, not generated from within. The search for profit and cost-savings tends to be a more obvious instigating factor of technical change, and it might be argued that market-based mechanisms, a technology policy which is complemented by a regulatory framework, and a good corporate environmental management strategy, can better contribute to achieve that aim.

There has been growing interest in the use of market-based mechanisms, whereby the polluter is charged for destructive use by estimating the damage caused. An important justification for the use of market-based incentives is that they allow companies greater freedom to choose how best to attain a given environmental standard.[1] By remedying market failures or creating new markets (rather than by substituting government regulations for imperfectly functioning markets), it has been argued that market-based incentives may permit more economically efficient solutions to environmental problems. Two categories of incentives exist.[2,3] One set, based on prices, includes a variety of pollution taxes, emission charges, product charges, and deposit-refund systems. Another set is quantity-based, and includes tradeable pollution rights or marketable pollution permits. The most common of these measures relates to posting bonds up-front for the rehabilitation of mines on closure. This is standard practice now in Canada and Malaysia. There are also discussions taking place about a mercury tax in Brazil (see Cleary and Thornton's article earlier in this volume for further details) and a cyanide tax in the USA. Currently, no government has designed a systematic set of incentives for industry to innovate and develop new environmental technology.

There are two further areas where policy approaches can contribute to improved environmental management practices. First, the increasing conditionality of private, bilateral, and multilateral credit, which frequently requires both prior environmental impact assessment and the use of best practice environmental control technologies in new mineral projects. A growing number of donor agencies, in Germany, Canada, Finland, and Japan, for example, are also concerned with training in environmental management. Second, the attempts by some governments, particularly Canada, to promote research and development activities (jointly and within industry and academic institutions) to determine toxicity from mining pollution and clean-up solutions. For example, Canada has extensive government-funded research and development programmes to promote the abatement of acid mine drainage and of SO_2 emissions. There is considerable scope for expanding these approaches, as is argued in Section 4.

Environmental regulations designed specifically for mining and mineral processing have, until recently, been uncommon in developing countries, although most countries now have in place basic standards for water quality and,

[1] OECD, 'Environmental Policy: How to Apply Economic Instruments', Paris, 1991.

[2] D.C. O'Connor, 'Market based incentives', in 'Environmental Degradation from Mining and Mineral Processing in Developing Countries: Corporate Responses and National Policies', discussion document, ed. A. Warhurst, SPRU, 1991, Section 2, Ch. 5, pp. 189.

[3] A. Warhurst, 'Environmental Management', Chapter 7, in 'Mining and the Environment: the Berlin Guidelines', Mining Journal Books, 1992.

less commonly, air quality. A few developing countries have recently adopted extensive regulatory frameworks—sometimes replicas of US models. This, for example, has been the case in Chile and, to a lesser extent, in Brazil. This growing concern about environmental degradation is occurring during a period of rapid liberalization in developing countries,[4] which finds expression in new policies to promote foreign investment, privatization schemes, and the availability of loan capital. These conditions also influence the regulatory regime of developing countries. Should the developing country pose less onerous environmental burdens on the potential investor to improve the terms of the investment by implying lower compliance costs or a greater assumption by the state of the environmental costs associated with mineral development projects? Should agreements be signed which release new investors from any liability for environmental damage caused by previous mine owners under less-restrictive regulatory regimes? Or will a clear and strict regulatory regime be more likely to facilitate credit flows from increasingly more environmentally conscious lending agencies? Developing countries, desperate for investment in their stricken mineral sectors, will need to determine what the market can stand and how such terms can be structured to reduce to the minimum the risk premium the investor will seek for a given tax or regulatory burden.

It is worth noting that surveys by Johnson[5] and Eggert[6] imply that environmental policy has not been a major factor in determining the investment strategies of international mining companies. However, more recently the industry press has been citing environmental regulations in Canada and the USA as a major factor causing the cancellation and delay of potentially large investment projects[7] and contributing to the shut-down of several mines. For example, in 1989, the Bharat Aluminium Company announced the closure of its bauxite mining project in the Gandhamardham Hills, Orissa State in India, because of strong environmental opposition by the local population.[8] Other projects such as Phelps Dodge, Copper Basin in Yavapai County (USA) have been withdrawn due to delays and excessive costs involved in project approval, while in 1991 the Kennecott Flambeau Mining Company finally received planning permits after twenty years of negotiation for the Grant Copper Mine in Wisconsin, which will operate for only six years.

However, environmental regulation alone is unlikely to solve environmental problems in developing countries due to endemic production inefficiencies. In particular, the approach of state-owned enterprises towards the environment reflects inefficient operating regimes, excess capacity, breakdowns and shut-downs, and poor management procedures, which contribute to worsen the polluting nature of effluents and emissions. Such inefficiencies make it very unlikely that environmental controls will be incorporated effectively.

[4] R. Brown and P. Daniel, 'Environmental issues in mining and petroleum contracts', IDS Bulletin, 1991, **22** (4).

[5] C.J. Johnson, 'Ranking countries for mineral exploration', Natural Resources Forum, August 1990, **14** (3), 178–186.

[6] R.G. Eggert, 'Exploration', in 'Public Policy and Competitiveness in the Nonferrous Metal Industries', eds. J.J. Landsber, M.J. Peck, and T.E. Tilton, 1992.

[7] *Min. J. (London)*, 30 October, 1992.

[8] US Bureau of Mines, 'Bauxite, Alumina, Aluminium', Annual Report, 1989, p. 6.

Production inefficiency is endemic among many mining enterprises in developing countries, and problems of environmental degradation cannot be viewed independently of it. Moreover, obsolete technology is widely used without the modern necessary environmental controls and safeguards. For example, new concentrators and roasting plants tend to be totally computerized. Automatic ore assaying techniques give an extremely accurate picture of the chemical composition of the ore feed which has implications for the fine-tuning of pressure, heat, cooling, and specific environmental control systems. This, in turn, will facilitate the accurate prediction and monitoring of emissions. However, where these controls are missing and, in particular, where ore feeds are of variable composition (in terms of the sulfur, lead, and arsenic content) the pollutant content of emissions also varies. The inefficient use of energy and poor energy conservation practices also result indirectly in increases of environmental pollution through the excessive burning of fossil fuels. This is particularly the case in poorly lagged roasters and inefficiently operated flotation units and smelters which are very intensive in energy use. It might be further argued that command and control regulatory instruments are unlikely to result in a reduction of pollution since they cannot affect the capacity to implement technical change of a debt-ridden, obsolete, and stricken mining enterprise in the developing country context. Such a company might find it preferable to risk not being detected or convicted, to pay a fine, or to mask its emission levels, rather than face bankruptcy through investing in radical technical change.

In addition to the problems of inefficient production, there are further reasons as to why environmental regulations—particularly those of the 'command and control' and incremental 'paper tiger' nature[9]—do not improve environmental management, particularly in developing countries. These are discussed below.

Environmental regulations tend to be of the blanket-type which specify maximum levels of emitted substances, minimum levels of environmental quality, and best available technology standards. They do not tend to reflect the propensity of a particular operation to pollute, which in part depends on local site-specific conditions (geology, geography, and climate) as well as economic, infrastructure, and technology-related constraints. In a desert, tailings dams need not be as highly specified as in rainy climates; dust regulations may need to vary depending on topography, precipitation, and prevalent winds; the sub-strata of leach ponds might need to be of different composition, strength, and depth, depending on local geology or the existence of an impermeable level of clay. Since developing country regulations are often copied directly from the statute books of the industrialized countries (for example, there are instances in both the cases of Chile and Brazil) whose regulations are adapted to suit their circumstances, they may not be appropriate for the site-specific characteristics of mines in either tropical regions or deserts. They may result in unnecessary and costly adaptations on the one hand, or the lack of necessary control, on the other.

Command and control environmental regulations require intensive monitoring to ensure that they are enforced. However, the small and medium mine sector

---

[9] T. Panayotou, Q. Leepowpanth, and D. Intarapravich, 'Mining, Environment and Sustainable Land Use: Meeting the Challenge', Synthesis Paper No. 2, The 1990 TDRI (Thailand Development Research Institute), Year-End Conference, Jomtien, 8–9 December, 1990.

accounts for at least 25% of mineral production in many countries. Although these mines are individually relatively small polluters, collectively they account for a disproportionately large share. These mines are often located high in the Andes or in remote tropical rain forests and are almost impossible to monitor systematically. Indeed, as regulation becomes more sophisticated, such monitoring requires skills and human resources far beyond the technological and managerial capabilities of many developing countries and frequently beyond their budgets. Understanding the diverse range of toxicity and engineering issues behind regulatory aims also poses challenges even in the industrialized countries. The most knowledgeable regulators are often head-hunted by the mining companies.[10]

In a recent interview in Brazil, a spokesperson for one of the companies said that they were requested to monitor themselves and send effluent samples at intervals to an independent laboratory and to report any abnormal results. The State regulatory agency did not have the skills required to monitor the operation itself. Indeed, the skilled people involved in the environmental agencies tend to live and work in capital cities and infrequently travel into the politically dangerous and inhospitable mining regions. Moreover, the enforcement of command and control regulations depends on a system which admonishes with imprisonment and fines. This, in turn, requires a legal structure and judicial system far beyond the capacity of most developing countries. Compliance is also limited since fines are generally a fraction of the costs involved in remedial treatment and abatement technology. They are also only payable if the polluter is detected, and if convicted. Inflation and local currency devaluation, which are endemic in the developing country context, also eat into the value of such fines. The costs of environmental regulation enforcement are generally hidden from the public eye and regulatory agencies are not generally accountable as such. However, since different site-specific mining contexts often require individual regulation, perhaps for permit approval, this provides opportunities for bribery which is endemic in bureaucracies and industry in many developing countries.

Indeed the regulatory system does not demand that efforts are made to deal with the cause of environmental pollution once and for all. It simply deals with the symptoms—once they are reported. Even though there is a theoretical threat of mine closures due to non-compliance, most foreign mining companies know that their developing country host can least afford to lose the foreign exchange earnings from their activities. Therefore the risk of closure due to environmental non-compliance of this type is considered relatively low. Pollution rarely produces a one-off disaster—rather it is a constant crisis.

Environmental regulations often emerge contradicted by other economic and industrial policies. For example, several countries with tropical forests have recently introduced policies aimed at their conservation. At the same time countries such as Brazil, Ecuador, and Colombia have parallel economic policies to promote industrial investment, especially by foreign firms, in these remote areas. The example of the Government of Ecuador authorizing RTZ's mining investment in one of its national parks is one such case—it resulted in the latter company withdrawing to avoid controversy over the issue. Forest conservation

10 EPA, 1991.

policies were also in place in Brazil in line with EV and World Bank loan conditions. However, despite the existence of these policies, smelters at Carajas, in Brazil, were fuelled by large amounts of charcoal from neighbouring forests.

Another recently discussed example of potentially contradictory policy issues revolves around the international Basel convention which restricts the inter-country transportation of toxic wastes. Since certain scrap metals fit theoretically into this category on account of their heavy metal content, this would undoubtedly restrict trade in scrap metal and metal recycling.[11] This is considered to contradict many of the new intentions of European and American governments to encourage the recycling of metal-containing materials at the expense of new primary production.

Command and control regulation tends to identify and deal with symptoms (pollution) of environmental mismanagement rather than causes (production inefficiency, human resource constraints, lack of technology, and lack of capital). It is also add-on and incremental in nature. Therefore, there is a tendency for it to emphasize end-of-pipe, add-on, and capital-intensive solutions (*e.g.* smelter scrubbers, mine water treatment plants, dust precipitators, *etc.*) for existing technology and work practices rather than promote alternative environmental management systems or technological innovation. Regulation may also, to a certain extent, presuppose a static technology. If regulation is incremental, technical change may be incremental, involving the addition of numerous new controls at relatively greater cost and with more overall resultant degradation than if a new, more radical change had been introduced in the first place. It may also oblige specific reductions in pollution, regardless of cost or local context. For example, regulation will refer to the chemical composition of an effluent in isolation from how that discharge rate and pattern may be influenced by natural site-specific precipitation, evaporation, or soil and geological conditions. In turn, this regulatory approach may get a more uncooperative response from industry which sees the rules always changing and their cost implications increasing. Furthermore, such regulation ignores the human resource elements of sound environmental management by emphasizing a specific pollution control technology rather than training, managerial approaches, and information diffusion.

3 Technical Change and Corporate Environmental Trajectories

Enterprise responses to environmental pressures have been characteristically slow, and reflect the regulatory regimes and public climate of either their home country or foreign countries of operation. Their response has also depended on the nature of their operations in terms of first, the mineral involved; second, the level of integration of mining and processing activities; third, the stage in the investment and operations cycle which its mineral projects have reached; and fourth, the internal economic and technological dynamism of the company, *i.e.* whether it has the financial, technical, and managerial capabilities to be an innovator or not.

---

[11] OECD, 'Environmental Concerns Related to Commodities: Scrap Recovery and Recycling of Non-Ferrous Metals and Environmental Policy Instruments: The 1989 Basel Convention', Paris, 15 October, 1990.

After a period of using rather 'static' technology, the mining and mineral processing industry is going through a phase of technical change as dynamic companies are innovating by developing new smelting and teaching technologies to escape economic as well as environmental constraints. Rapidly evolving environmental regulatory frameworks in the industrialized countries and the prospects of their application, reinforced by credit conditionality in the developing countries, are stimulating this trend. Changed technological and environmental behaviour in this context is evident particularly in the large North American and Australian mining companies, but is becoming increasingly apparent in developing country-based companies operating in, for example, Chile, Brazil, and Ghana. However, it seems to be the new operators and dynamic private companies which are changing their environmental behaviour, while both state-owned enterprises and small-scale mining groups in developing countries continue, with some exceptions, to face constraints regarding their capacity to change environmentally damaging practices.

It is somewhat inevitable that only those companies which are dynamic and with new project development plans are in a position to invest in the research and development required to develop more environmentally sound alternatives, or to raise the capital to acquire them from technology suppliers. Nonetheless, after a long period of only conservative and incremental technical change, alternative process routes for mineral production are being developed which emerge as being more economically efficient as well as environmentally less hazardous. Furthermore, companies are beginning to sell their technologies, preferring to commercialize their innovations to recoup their research and development costs rather than sell obsolete technology and risk shareholders' scorn or retrospective penalties as environmental regulations are increasingly enforced by the developing countries. Some of those companies have pushed technology even beyond the bounds of existing regulations and as a consequence are seeking to increase regulation—particularly on a worldwide scale—because they can meet the requirements and use their new environmentally sound technologies to their competitive advantage.

There is evidence that improving the environmental management of a mining operation may not necessarily be detrimental to economic performance, and in some cases may even be of economic benefit. Furthermore, because environmental regulation is here to stay and bound to become more widely adopted, more stringent, and better enforced, who wins in the division of shares in the metals market will not be those companies that avoid environmental control (only later to be forced to internalize the high cost of having done so), but will be those companies that were ahead of the game, those that played a role in changing the industry's production 'parameters', and those that used their innovative capabilities to their competitive advantage.

The 'environmental trajectories' that different mining enterprises might take in response to environmental and market pressures are categorized in Figure 1. This diagram could be a planning tool for both companies and governments. It can help to evaluate the environmental and economic implications of applying different policies on corporate development.

The average mineral enterprise is competitive (*i.e.* to the left of the threshold of

Figure 1 Corporate environmental trajectories

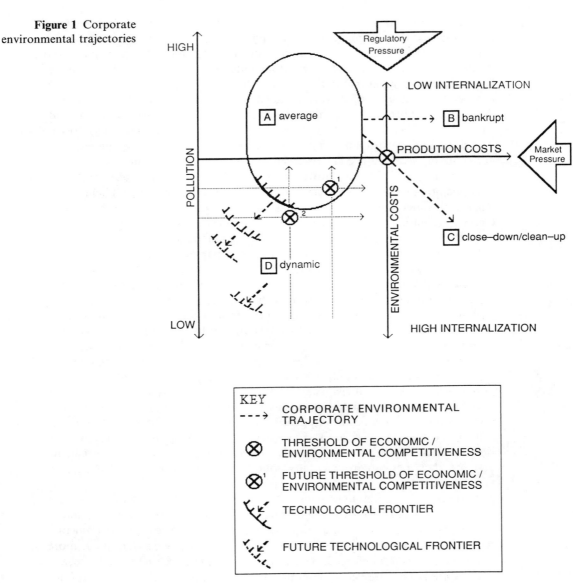

economic competitiveness, X), although to a greater or lesser extent these enterprises produce environmental pollution and to a greater or lesser extent they have internalized the cost of the environmental degradation associated with their metal production, in response to the regulatory regime they are working within. (The threshold of 'environmental competitiveness' for a given regulatory context is also X and company operations in compliance have environmental trajectories in the quadrants below the horizontal axis.) However, market pressures—mainly a real decline in metal prices—combined with their economic inefficiencies, mean some of these companies are going bankrupt (a trajectory towards quadrant **B**).

They will leave a legacy of environmental pollution behind, and as in the case, for example, of COMIBOL (in Bolivia) and Carnon (in the UK), the burden of clean-up will fall on the state and society. Other companies will respond by innovating; moving into quadrant D; building into the new generation of technology both improved economic and environmental efficiencies (protecting themselves in the process from having to undertake relatively more costly add-on, incremental technical change and rehabilitation at later stages in their operation). Indeed, freed from the incumbent costs of retrofitting sunken investments, greenfield plants, in particular, display new levels of dynamism—the latest best-practice technology incorporates both improved economic and environmental efficiencies.

Nonetheless, there exists a growing group of companies which, if obliged to 'add-on' environmental controls in line with new regulations, would have to close down since the cost of the controls and clean-up required would render their operations uneconomic. The environmental trajectories of these companies is towards quadrant C. Currently, such examples are few, and it is difficult to differentiate between purely environmental factors and the range of other reasons as to why a company's cost curve starts increasing. However, as Figure 1 shows, that group would be expected to grow in number, since market and regulatory pressures combined will lower the threshold of economic and environmental competitiveness such that the average company will only survive in the new regime if it innovates. Even the previously dynamic companies will need to keep their environmental trajectories moving ahead of the encroaching threshold of economic/environmental competitiveness (X^1 and X^2).

Moreover, this implies a serious constraint on the regulatory process for two main reasons, which indeed distinguish mining companies from their manufacturing counterparts.

First, an implied close-down due to regulatory burden does not signal the end of environmental degradation. Pollution in metals production is not all end-of-pipe. Rather it heralds a new era—decommissioning, clean-up, and rehabilitation all pose significant environmental costs.

Second, in very few countries are bankrupt operators liable for the clean-up of their 'sins of the past'. The USA with its 'Superfund' liability laws, is an exception. Therefore, by moving the threshold of economic and environmental competitiveness, the overall extent of environmental degradation (particularly that without liability) increases. The policy challenge of the environmental imperative is therefore how to keep firms sufficiently dynamic to be able to afford to clean up their pollution and generate economic wealth.

Cleaning-up 'Sins of the Past' and Some Environmental Effects of Current Regulation

It is inevitable that those companies with long mining histories and extensive sunken investments in conventional mining and smelting facilities face the greatest technical, and therefore economic, challenges in cleaning-up their past facilities and reducing pollution from their ongoing operations. For example, some companies in central and south-west USA have found that dumps from past lead and copper mining operations have now created such serious acid mine drainage and toxic seepage that the government has placed them on its 'Superfund List' which obliges multi-million dollar sums being spent on their clean-up. The government then targets previous owners of the mine, often the richest, making them liable. If a company has already closed down a mine and written off the investment and perhaps is struggling in the current economic climate to manage a new project, it is clear that the costs of such a Superfund indictment, and the legal costs involved in answering it, can be quite crippling. For example, the Smuggler Mountain lead mining site in Colorado has a serious acid mine drainage and toxic seepage problem.[12] Its old lead and cadmium mine workings have apparently contaminated soils and ground-water in neighbouring residential areas, requiring a major clean-up operation, the secure repositioning of the toxic waste, and the establishment of monitoring mechanisms and pollution controls to prohibit further contamination. The cost of this project as estimated by EPA is currently US $4.2 million. Another Superfund listed mine site is the Silver Bow Creek Site in Butte Area, Montana, which for over one hundred years has been mined for silver, copper, gold, and zinc, resulting in severe water and soil contamination and the disruption of local ground and surface drainage water patterns. Currently, ground water is flooding the mine, becoming highly acidified in the process, and it is absorbing high concentrations of iron, manganese, arsenic, lead, cadmium, copper, zinc, and sulfate. This toxic seepage is currently threatening Silver Bow Creek, a major river in the region. The clean-up and remedial action is extensive and involves detailed diagnostic analysis and monitoring, water and tailings containment, water treatment, and soil treatment. Major mining companies and individuals, all past owners, are implicated, including Atlantic Richfield, AR Montana Corporation, and ASARCO. The clean-up cost is estimated to be in the region of several million dollars, to be confirmed once the precise plan of remedial action is determined. Another Superfund listed old mining site is that of Gregory Tailings in Colorado.[13] It was a gold mine, exploited during the late nineteenth century. Waste was placed in inadequate tailings dams and resultant leakages have contaminated local water supplies and soils with acidic waters containing copper, zinc, nickel, cadmium, and arsenic. The cost of the clean-up and water treatment has yet to be confirmed, but the strengthening of the tailings dam is estimated at over half a million dollars.

Although there may be economic opportunities associated with clean-up operations, such as the recovery of extra metal values from acid mine drainage, the commercialization of innovative water treatment methods, or the innovative use of tailings material, these may not be recouped by the mining company itself. Furthermore, companies which previously have had no links to the facility may be nervous of getting involved in case any liability is passed on to them.

This is one reason why some mining companies, which need to clean up their past operations, object to such retrospective regulation and suggest that such restrictions and control threaten their existence. An interesting and illustrative case is that concerning the respective responsibility of both government and industry for the management of mine closure and rehabilitation of old tin,

[12] EPA, 'Soil Clean-up of Smuggler Mountain Site', March, 1989.
[13] EPA, 'Gregory Tailings', July, 1986.

copper, and silver mining and smelting operations in Cornwall, UK. The observations that mine closure and rehabilitation were proceeding very inefficiently in the absence of an adequate regulatory framework have been borne out by the recent flooding of polluted acid mine water from the Wheal Jane mine near Truro, Cornwall (Warhurst, Mining and Environment Research Network Newsletter No. 2, April 1992). This highly acidic cocktail of dissolved metals, including copper, lead, cadmium, tin, and arsenic, entered the Carnon and Fal rivers at a rate of 2–4 million gallons a day and has spread throughout the surrounding estuary and coastal areas. The tourist industry and fisheries have been threatened and local well-water supplies destroyed.

Many have argued that legal responsibility for the mine discharge rests with the current owners, Carnon Consolidated (a management buy-out in 1990 from RTZ, which had bought it seven years previously from Consolidated Goldfields). However, the National Rivers Authority (NRA), the relevant UK regulatory body, publicly admitted that they were aware six months previously that this disaster could happen, but were unable to prevent it since their policy remit did not cover preventative action. The disaster occurred following the withdrawal of a government grant to the Wheal Jane mine which had always required pumping to prevent it from flooding. This meant that plans had to be shelved for turning the site into a golf course and leisure centre to help fund the pumping and maintenance costs at the mine of £100 000 per month. The consequent lack of finance forced Carnon Consolidated to turn off all the pumps on January 4, 1992. Since they had officially abandoned the mine, they denied responsibility for any ensuing flooding and pollution, a claim complicated by the fact that numerous underground shafts from other abandoned mines in the region also provided conduits for the polluted floodwater. The NRA had stated its intention to prosecute Carnon, but loopholes in the 1991 Water Resources Act meant there was no UK regulation to ensure that previous mine owners bear the financial liability for clean-up measures (unlike the Superfund legislation in the USA). Abandoned mines are specifically exempted from clean-up liability. Furthermore, the NRA itself has no remit or budget to treat pollution, particularly on this scale.

The NRA has opened an old dam to hold and treat the contaminated water. It is also analysing the potential of biotechnology to assist in the clean-up. For example, certain microbes bred in a slurry of cattle excrement have been shown to be metal-absorbing. Similarly, the creation of wetlands containing plants whose roots absorb metals is another possible long-term solution (see Cairns, Jr. and Atkinson's article for detail). However, such solutions are based on piecemeal research being undertaken in this area in different research institutions (*e.g.* the Colorado School of Mines in the USA, CANMET in Canada, and CETEM in Brazil) and none of these techniques has yet been proven commercially. The NRA is approaching the UK government to help pay for the clean-up, which may take decades and could cost over £1 million. The ultimate financial responsibility for mine clean-up in the UK may therefore lie with the tax-payer.

Environmentalists in the UK are pressing for the law to be changed so that companies which abandon mines are liable for any resulting pollution. However, it has been suggested that such a change would make life difficult for British Coal. The NRA has already analysed pollution from rising waters in old coal mine workings which are polluting rivers in South Wales and Yorkshire. They have established that the capital costs of treating the ten worst cases in Yorkshire are estimated at more than £10 million, and indeed as long ago as 1981 the Royal Commission on coal and the environment (the Flowers Report) recommended that the costs of remedial action for existing mines abandoned by the National Coal Board are met by central government. However, the tightening of UK laws to ensure that the bills for such pollution are paid by previous owners could have serious implications for the current government's plans to privatize the coal industry.

Similar pollution liability issues are also being faced by governments in developing countries currently engaged in privatizing their state mines—Bolivia, Peru, and Chile are cases in point. In Peru, for example, the legacy of past pollution, particularly from toxic tailings along the river below CENTROMIN's La Oroya and Cerro Pasco facilities, was preventing the government from selling those enterprises since the cost of clean-up rendered the investments uneconomic and unattractive to foreign capital. It was therefore agreed that the investment contract for buying these operations would protect the foreign partner from liability for previous environmental damage. They would start with a 'clean slate', as it were, with generous lag-times regarding the introduction of new environmental controls to reduce ongoing pollution. This means that the economic burden for clean-up falls on the state but, in developing countries, that means on society. Where capital is scarce cleaning up pollution problems from past decades affecting remote rural communities is of low priority.

This poses a policy dilemma. If the government does not waive liability for past pollution, the privatization schemes, vital to the future of the economy, will not succeed since foreign partners would not be interested relative to other available investment opportunities in mining. Moreover, international companies are particularly wary of falling prey to new retrospective liability laws and punitive tariffs in their import markets, particularly given the enormity of the clean-up involved. The responsibility for clean-up therefore lies formally with the state. Should loan-conditionality put pressure on the government to clean-up using precious capital resources? Should clean-up funds be established and incentives provided to prompt local industry to develop technical solutions? Should aid programmes provide technical assistance and training in clean-up? Should new investors be taxed for old pollution? Clearly the optimal outcome would be for government and donors combined to provide incentive programmes for local firms to seize the commercial opportunities available. Nuñez[14] clearly documents the extensive range of local capabilities which could be harnessed.

While 'Superfund' in the US may theoretically be successful, given that one is targeting local investors and traceable companies, it may be more difficult to target and litigate against previous mine owners (prior to nationalization) in developing countries. Searching out the foreign investors responsible for the many old and abandoned mines may be difficult since most have long since returned home, and local miners have limited resources, which makes the task of determining liability and enforcing clean-up a daunting one.

[14] A. Nuñez, 'Environmental Management in a Heterogenous Mining Industry—The Case of Peru', Paper presented at the second workshop to the Mining and Environment Research Network, Sussex, September 1992.

Companies which are being forced through environmental pressures to deal with pollution problems in their existing operations have been observed to react both defensively and in innovative fashion, depending on the challenges posed. For example, depending on the level of enforcement of the regulatory regime, some mining companies, particularly those operating in the developing country context, may prefer to pay financial penalties and fines for affected water and air quality. These may amount to less than the cost and effect involved in remedial action, such as water treatment, and considerably less than the costs involved in innovation or the incorporation of pollution controls. In some instances, as discussed above, the state pays those remedial costs itself and it is clear that it may be subsidizing the profit of foreign mining companies at the expense of environmental degradation. Sometimes that trade-off is influenced by the state's absolute dependence on the foreign company as a source of foreign exchange and government revenue.

Indeed, a number of mining companies perceive environmental regulation as imposing a cost burden on their operations that threatens their profitability. They may then enter into negotiations with the state to arrive at a 'stay of execution' or to devise a plan for implementing controls. However, as regulation becomes increasingly strict, backed up by more sophisticated monitoring devices and data processing, companies are being pushed to take remedial action in both the industrialized and developing countries. Data from the USA suggests trends that may be followed elsewhere. According to a US Congressional Report, sulfur dioxide emission controls have resulted in 'substantial capital expenditure' for US copper smelters and increased operating costs due to 'add-on' acid plants.[15] Present levels of environmental control entail capital and operating costs of between 10 and 15 cents per pound (*i.e.* 27–40¢ kg^{-1}) of copper. However, the USA has lost substantial smelting capacity. It has been reported that eight out of sixteen smelters operating in the US in the late 1970s have closed permanently, 'most because the capital investment to meet regulations was unwarranted given current and anticipated market conditions'.[16]

Moreover, the evidence from studies in the USA shows that environmental compliance does not distort significantly the economics of new mineral projects, but does place a considerable cost burden on ongoing facilities for either retrofitting or clean-up on mine and plant closure. The US Bureau of Mines has estimated direct 'environmental' operating costs for smelting facilities with emission controls. Retrofit capital costs were estimated to be of the order of $150 million per facility or 5.6 cents per pound (15¢ kg^{-1}) of copper produced.[17] According to Coppel[15] the overall cost penalty, including capital invested, to the producer for implementing the new smelter and sulfur dioxide capture facilities was estimated to be 7.5 cents per pound (20¢ kg^{-1}) after deductions of 1.3 cents per pound (3.5¢ kg^{-1}) of acid credit. The operating costs for individual smelters ranged from 10 to 15 cents per pound (27–40¢ kg^{-1}) of copper, and the average

[15] N. Coppel, 'Worldwide minerals and metals investment and the environment, 1980–1992', unpublished report for RTZ Corporation Plc, August, 1992.

[16] Copper Technology and Competitiveness, Office of Technology Assessment, Congress of the US, 1988, p. 16.

[17] US Bureau of Mines, 'Copper', Annual Report, 1990, p. 28.

operating cost in 1987 was 12.3 cents per pound (33¢ kg^{-1}). Of this amount, 26% or 3.2 cents per pound (8.6¢ kg^{-1}) was calculated by the US Bureau of Mines to be the cost burden of compliance with environmental, health, and safety regulations.

It is highly like that regulation in developing countries will follow a similar pattern. An interesting example is the ALCAN bauxite mine and alumina plant in Jamaica. Foreseeing impending environmental regulation and responding to public concerns in its home country of Canada, ALCAN supported a local university department to develop an innovative solution to the disposal of the red mud sludge from its bauxite mining operations. Previously, the sludge was dumped in a large catchment pond, but toxic seepages into surrounding soils and groundwater had been reported. The university developed a process called red-mud stacking which involved sun-drying of much of the moisture content of the sludge and stacking of the material into much less obtrusive piles, which indeed as bricks may have further use within the plant site.[18] A similar technology was introduced by ALCAN at the Vaudrevil alumina plant in Quebec in 1989.[19]

However, this technology neither seems to offer a solution to the toxic seepage of previously dumped slurries nor is it a solution to pollution *per se*. It appears to replace water pollution by dust pollution which is less stringently regulated against. Moreover, a change in the production process to facilitate the recovery of caustic soda from the 'mined' dry mud stacks means that a greater amount of that chemical is discharged than in the previous disposal method. This means that the dust pollution, plus overflows from those parts of the dry mud stacks which become water-logged during tropical rain showers, may cause a greater toxic hazard than previous low level seepages.

Dynamic Innovators—Technical Change to Improve Environmental Management

Although some mining companies resist the application of environmental regulation to their existing operations, a growing number of dynamic innovative companies are making new investments in environmental management because they see an evolution toward stricter environmental regulation. Free of the encumbrance of sunken investments in pollutant-producing obsolete technology or with significant resources for research, development, and technology acquisition, they have chosen either to develop more environmentally sound alternatives or to select new improved technologies from mining equipment suppliers, who are themselves busy innovating. Increasingly, these new investment projects are incorporating both improved economic and environmental efficiencies into their new production processes, not just in terms of new plant or items of technology, but also through the use of improved environmental management practices. Some examples of these are discussed below in three categories: smelter emissions, gold extraction, and waste management.

Smelter emissions.   **Inco Ltd.**: At one time one of the world's highest cost nickel producers, Inco was until recently the greatest single point source of environmental pollution in North America. This was due to its aged and inefficient reverberatory

---

[18] D. R. Kelly, Personal Communication, Alcan International Ltd., Montreal, 1990.
[19] US Bureau of Mines, 'Bauxite, Alumina, Aluminium', Annual Report, 1989, p. 6.

furnace smelter technology which spewed out excessive tonnages of SO_2 emissions. Inco Ltd. had reached the limit of improving the efficiency of this obsolete technology through incremental technical change at the same time as the Ontario Ministry of the Environment began an intensive SO_2 abatement programme to control acid rain. These factors prompted Inco to invest over 3000 million Canadian dollars (C$) in a massive research, development, and technological innovation programme.[20]

Under the Canadian acid rain control programme, Inco is required to reduce SO_2 emissions from its Sudbury smelter complex from its current level of 685 000 to 265 000 tonnes per year by 1994: a 60% reduction. To achieve this, Inco plans to spend C$69 million to modernize milling and concentrating operations and C$425 million for smelter SO_2 abatement. The modernization process will include replacement of its reverberatory furnaces with a new innovative oxygen flash smelter, a new sulfuric acid recovery plant, and an additional oxygen plant. By incorporating two of these flash smelters the company plans to reduce emissions by over 100 000 tonnes per year in 1992, and by 1994 to achieve the government target levels of 175 000 tonnes per year. Other environmental benefits include a cleaner, safer workplace environment.[21]

Inco is now one of the world's lowest cost nickel producers and again, like other dynamic companies that are responding to environmental regulation through innovation, Inco is seeking to recoup research and development costs through an aggressive licensing effort in other copper and nickel processing countries. More than 12% of Inco's capital spending during the last ten years has been for environmental concerns.[15]

Kennecot—Utah, USA: A new smelter project has recently been launched by Kennecott Corporation (RTZ) with the dual aims of setting a new standard for the cleanest smelter worldwide and improved cost efficiencies in processing its ore. Advantages include the capture of 99.9% of sulfur off-gases (previous levels were 93%). Sulfur dioxide emissions will be reduced to a new world best-practice level of approximately 200 pounds (*i.e.* 75 kg) per hour, less than one twentieth of the 4600 pounds (1716 kg) per hour permissible level for Utah's current clean-air plan. The investment is US $880 million, resulting in 3300 new construction jobs and the investment of US $480 million in local companies through project development contracts.

The proposed Garfield smelter will expand the concentrate processing capability to the level of mine output (about 1 million tonnes of copper concentrate per year) at more than half previous operating costs. It represents the first-time application of oxygen flash technology to the conversion of copper matte to blister (details are based on an excerpt from a Kennecott Corporation press release, March 11, 1992). The two-step copper smelting process consists of smelting furnaces which separate the copper from iron and other impurities in a molten bath, followed by converting furnaces where sulfur is removed from the molten copper. A new technology known as flash converting will then be utilized in the second step of the process at the new smelter. This unique technology was developed by Kennecott in co-operation with Outokumpu Oy, a Finnish

[20] R. Aitken, Personal Communication, Inco Ltd., 1990.
[21] *Min. J.* (*London*), 23 February, 1990.

company and a leader in the supply of smelting technology. Essentially the new technology eliminates the open air transfer of molten metal and substitutes a totally enclosed process of producing molten metal. 'Flash converting' has two basic effects: first, it allows for a larger capture of gases than the current open-air process; second, it allows the smelter's primary pollution control device—the acid plant—to operate more efficiently. The smelter will include double-contact acid plant technology.

There will be other environmental benefits from the new smelter as well. Water usage will be reduced by a factor of four through an extensive recycling plan. Pollution prevention, workplace safety, and hygiene and waste minimization will be incorporated into all aspects of the design. In addition, the smelter will generate 85% of its own electrical energy by recovering energy as steam from the furnace gases and emission control equipment. This eliminates the need to burn additional fossil fuel to provide power. The new facility will require only 25% of the electrical power and natural gas now used per tonne of copper produced.

The copper refinery's planned modernization and expansion modifications include major electrical system changes, material handling system improvements, and new electro-refining cells. In addition, a new state-of-the-art precious metals refinery will be built. The refinery will be able to process the entire output of the new smelter.

Gold Extraction. **Homestake's McLaughlin Gold Mine, California**: The McLaughlin gold mine, opened in 1988, is perhaps the best example of a new mine and processing facility which has been designed, constructed, and operated from the outset within the bounds of what is probably the world's strictest environmental regime. Environmental efficiency is built into every aspect of the gold mining process, in terms of innovative process design criteria, fail-safe tailings and waste disposal systems, and extensive ongoing mine rehabilitation and environmental monitoring systems. The mining operation therefore combines innovative technologies with 'best practices' in environmental management. The most interesting conclusions drawn by the author from site visits and discussions with the firm's environmental officers is that most of these environmental management initiatives have not resulted in any substantial extra cost, and indeed many of these procedures have apparently improved the efficiency of the mine, affecting positively the economics of the overall operation.

For example, before the mining operation began, an extensive environmental impact analysis and survey were undertaken. All plant and animal species were identified and relocated, ready for rehabilitation on the completion of mining operations. The survey also measured in detail prior air, soil, and water quality characteristics and flow patterns to provide the baseline for future monitoring programmes. Assaying was undertaken not just of the gold ore, but also of the different types of gangue material and waste, so that waste of different chemical compositions could be mined selectively and dumped in specific combinations to reduce its acid mine drainage generating capacity. Local climate conditions were evaluated to determine the frequency of water spraying needed to reduce dust, and evaporation rates were evaluated to control the water content and flood risk potential of tailings ponds. The tailings ponds themselves are constructed on specially layered impermeable natural and artificial filters, with high banking to prevent overflow, and with secondary impermeable collecting ponds in the rare case of flooding.

Unlike other mining projects where rehabilitation is seen as a costly task to be undertaken at the end of a mining operation, often when cash flows are lowest as ore grades decline, at Homestake rehabilitation began immediately and is an ongoing activity. Not only does this serve to spread expenditure more evenly over the life of the mine, but it enables the more efficient utilization of truck and earth moving capacity as well as of relevant construction personnel. This means that as soon as work piles have reached a certain pre-determined dimension, soils (previously stripped from the mine area and stored) are laid down and revegetation is begun. Although mining has been under way for only three years, extensive areas of overburden and waste have already been successfully revegetated—immediately reducing environmental degradation and negative visual impacts. In addition to these inbuilt environmental control mechanisms, Homestake Mining Company has sophisticated environmental monitoring procedures in place. This means that seepages, emission irregularities, wildlife effects, and vegetation effects can be detected and rectified immediately, which in the long-term reduces the risk of expensive shut-down in operations, costly court cases (*e.g.* if water toxicity results), and the need for treatment technologies.

Waste Treatment. In the minerals industry, considerable waste is produced in the form of overburden, marginal ore dumps, tailings, and slags. Much of the toxicity associated with that waste is a direct result of the loss of either expensive chemical reagents, or metal values. Currently, public policy has not taken up the challenge to promote and direct research and development in the area of waste reduction and treatment innovations. One interesting area is the application of biotechnology to waste treatment[22]

Water Treatment at Homestake's mine at Lead (S. Dakota, USA): Homestake Gold Mining Company turned regulatory pressure to clean up a cyanide seepage problem to its advantage. Its own research staff developed a proprietary biological technique to treat the effluent which led to the recovery of local fisheries and water quality in the mine's vicinity at Lead, North Dakota, USA.[23] It is now actively commercializing the technology which could be widely applied at other gold leaching plants.

Water Treatment at Exxon's Mine, Los Bronces (Chile): A mining project in Chile, Los Bronces, is to be expanded into one of the largest open-pit copper mines in the world and consequently required the stripping of very large tonnages of overburden and low-grade ore. Before mine development, the Chilean government warned Exxon that it would be imposing financial penalties for the water treatment costs on account of expected acid mine drainage from the overburden of low-grade ore dumps into the Mantaro River, the source of Santiago's drinking water. This threat became the economic justification for a

---

[22] A. Warhurst, 'Metals Biotechnology for Developing Countries and Case Studies from the Andean Group, Chile and Canada', Resources Policy, March, 1991, pp. 54–68.
[23] D. Crouch, Personal Communication, McLaughlin Mine, CA, 1990.

bacterial leaching project at the mine. Indeed, the feasibility of this bacterial leaching project was particularly illustrative of the profitability of leaching copper from waste at the same time as prohibiting otherwise naturally occurring pollution (acid mine drainage). Over a billion tonnes of waste and marginal ore below the 0.6% copper cut-off grade are expected to be dumped during the project's lifetime. The waste would have an average grade of 0.25% copper and would therefore contain a lucrative 2.5 million tonnes of metal worth approximately US 3.5×10^9, at 1985 prices.[24] The study demonstrated that with a 25% recovery, high quality cathode copper could be produced profitably, at 39 cents per pound (*i.e.* 105¢ kg^{-1}), by recycling mine and dump drainage waters through the dumps over a twenty-year period. This was shown to have the double advantage of both extracting extra copper and avoiding government charges for water treatment. At the same time both investment and operating costs were less than two-thirds of estimated costs for a conventional water treatment plant, which would not have had the benefit of generating saleable copper. The Los Bronces mine thus demonstrates the potential economical benefits of building environmental controls into mine development.

In conclusion, these few examples suggest that dynamic companies are not closing down, re-investing elsewhere, or exporting pollution to less restrictive developing countries; rather they are adapting to environmental regulatory pressures by innovation and by improving and commercializing their environmental practices at home and abroad.

4 Policy Implications for Mineral Producing Countries

Technical Change and the Environmental Trade-off

Evidence suggests that, at least during the 1980s, environmental policies have not been a major factor in determining where a mining company will target exploration and subsequent investment activities. Geological potential remains of primary importance, which is not to underestimate that in some cases the approval and permitting process is a major cost of compliance. (Johnson[5] ranked corporate criteria in selecting countries for exploration. See also Eggert.[6]) This would suggest that developing countries are not seen as pollution havens and that the industrialized countries' environmental regulations are not stifling new mining investment. Indeed, there are currently several new gold projects in the process of development in California, which has probably the world's strictest environmental regulatory regime. Although environmental policies may not negatively influence the investment activities of dynamic adaptive mining companies, the latter still seek to play a role in determining the detail and focus of relevant legislation so that new regulatory frameworks also reflect, as far as possible, their corporate interests. During preliminary fieldwork by the author in North America and Europe it became evident that this task was an important function of many of the companies' newly appointed environment vice presidents,

---

[24] A. Warhurst, 'Employment and Environmental Implications of Metals Biotechnology', World Employment Programme Research Working Paper, International Labour Organization (ILO), Geneva, March 1990, WEP 2–22/SP.207.

directors, and environmental affairs representatives. For example, the Environmental Vice President of Inco sits on Canada's high level Environment and Economy Committee, and the Environmental Director of Homestake Mining Corporation sits on the Environmental Committee of the American Mining Congress—which works closely with the US Environmental Protection Agency and lobbies government for tariffs on metal imports originating from countries with poor environmental performance.

If one understands the new environmental pressures being placed upon the mining industry in the industrialized countries in the context of hard-won survival following a prolonged period of low metal prices, which gave significant market advantages to their lower cost competitors in the developing countries, then it is possible to understand the recent lobbying by some firms for industry-wide international environmental standards. Although international standards may not pose too much of a problem for the economics of new mining projects in the developing countries, our analysis suggests there could be major costs incurred by any older ongoing operations. Controlling the latter's pollution problems would in most cases require major water treatment plants, strengthening and rebuilding tailings dams, add-on scrubbers, and dust precipitators, *etc*.

The imposition and strict regulation of international environmental standards could make some developing countries' mineral production uneconomic, thus swapping one social cost (environmental pollution) for another (unemployment, poverty, and indeed clean-up, given the absence of liability laws). This is not to dispute the need for improved environmental controls, particularly in the developing countries, but rather to show the complexity of the process by which the underlying power structure of the industry can help to determine the environmental agenda.

Most planned mines (including existing mine expansions) and available reserves are located in the developing countries. Furthermore, after a period of mineral production monopolized by state-owned mining companies (with some exceptions), many developing countries are now embarking upon a phase of liberalization and have promulgated a number of laws and incentives to promote foreign investment. In many cases those investments are being partly financed by credit which is conditional upon good environmental practice and prior environmental impact analysis. The upshot is that this trend in technical change may be to the benefit of the developing countries in that it may enable them to reduce the trade-off between higher environmental costs and lower production costs. This may mean that at least in the case of new mineral projects there may be a wider range of more environmentally sound and economically efficient technologies available to them.

Indeed, new flexible-scale, lower-cost, less-hazardous hydrometallurgical (leaching) alternatives to conventional smelting may further be to the advantage of developing countries, improving value-added from their mineral production. For example, processing right up to the stage of a final saleable metal product can be undertaken at the mine site—while in conventional process routes a smelter will require feed from at least ten large mines for full capacity utilization, and ore may have previously been exported to foreign smelters with consequent loss of by-products and entailing charges for the treatment of pollutant elements.

However, this new prospect of environmental security may have its own costs which require careful analysis. For example, depending how technology transfer agreements are drafted and managed, such new 'environmentally friendly' investment may herald indebtedness, bankruptcy of local equipment suppliers and engineering firms, and the loss of employment, *etc.*, reinforced by aid conditionality. On the other hand, smelting, concentration, and leaching innovations are being developed by international companies, such as Outokumpu, Mitsubishi, Kennecott, Inco, Cyprus Mines, and Homestake, which are adapted to new and prospective regulations in the industrialized countries. These trends may oblige developing countries for both economic and environmental reasons to export only semi-processed minerals or raw materials, reinforced by credit conditionality, new international regulatory agreements, and trade tariffs imposed on imports of metals produced not using a pre-determined 'Best Available Technology'.

Technology Policy for Environmental Management

Environmental behaviour correlates most closely with a company's capacity to innovate, rather than its size, origin, scale, and scope of operations or ownership structure. For example, government policy over time has resulted in a failure by state enterprises to re-invest capital in human resource development, repairs and preventative maintenance, research and development, and technology development. Managers became bureaucrats rather than entrepreneurs.[25] This factor, combined with cumulative inefficiencies, a poor waste management strategy leading to metal and reagent losses, and scarce resources, means that environmental mismanagement is endemic. It is a structural problem and one not readily solved by recourse to regulation, punitive tariffs, or even the simple act of purchasing an environmental control technology. The cases of COMIBOL (Bolivia) and MINEROPERU and CENTROMIN in Peru bear witness to this, as does the case of private companies such as Carnon Consolidated in the UK. However, CODELCO and ENAMI, the state enterprises of Chile, have invested in developing their innovative capabilities both within the industry and through historically close links with local research and development institutions and universities. Although new regulations currently pose a significant technological challenge to these companies, efforts are being made to develop the required human resources and to implement substantial technical change. For example, CODELCO is now at the forefront of metals biotechnology and has made considerable investments in new solvent–extraction/electrowinning technology. In addition, ENAMI is planning to replace its reverberatory furnaces with modern flash-smelting technology at an estimated cost of $300 million, 'largely motivated by the need for environmental improvements'.[15,26] Environmental degradation from small 'garimpeiro'-type operations is also related to the miners' incapacity to innovate through a lack of access to capital, technology, skills, and information. Scale further complicates the choice of optimal low-waste, high metals-recovery technology. With few exceptions, however, it is the private

[25] R. Jordan and A. Warhurst, 'The Bolivian Mining Crisis', Resources Policy, March, 1992.
[26] US Bureau of Mines, 'Copper', May–June, 1991.

sector which has so far shown itself to be most innovative and therefore most capable of improving environmental management. In several cases improved environmental management would have been brought about irrespective of regulation due to market pressures to introduce new, more efficient, low-waste technical change.

This is not an argument against regulation but rather to recommend a more sophisticated public policy approach through first, the definition of regulatory goals—something to aim at—and second, an informed technology policy to guide and stimulate those companies along the fastest, more efficient route to achieving those goals.

This technology policy could include a detailed technology-transfer strategy, tax relief on research and the training of engineers and managers in environmental technology, government grants for collaborative inter-industry and university–industry research projects, and information dissemination programmes regarding the moving technological and regulatory frontiers. It requires training for regulators so that they are informed disseminators of information about environmental technology. Finally, it requires the provision of new lines of credit—in banks and donor agencies—to promote investment in the development, commercialization, acquisition, and improvement of environmental control technologies.

Another factor which reinforces the need for trained and informed environmental regulators relates to their ability to determine corporate environmental trajectories as a response to regulation. Figure 1 provides a planning tool which would enable regulators to plot corporate environmental trajectories against changing thresholds of economic competitiveness and environmental compliance over time. For example, it is suggested that fast changing 'incremental' regulation would tend to promote add-on incremental technical change—pushing cost-curves upwards rather than down. As a consequence, and evidence of this is slowly emerging, many mining companies, particularly those with large sunken investments, are on a trajectory heading towards close-down and certainly reduced competitiveness. With the growth of market and regulatory pressures that group of companies heading towards bankruptcy, close-down, project delay, or cancellation is also likely to grow in number. At the same time, the clean-up problem posed by the closing-down of those companies would also be expected to grow in scale and severity, as evidence drawn for Peru[14,27] and Bolivia[28] demonstrates. Given the difficulties involved in retrospective clean-up legislation in terms of cost assessment, litigation, and technical complexity, it would seem desirable for government to avoid such a scenario. This could be done on the basis of sound prediction and planning, either through imposing a levy on operators as the mine nears exhaustion/abandonment (whichever is sooner) to cover clean-up and rehabilitation, or through promoting tax and other incentives for the required investment in clean-up. For new investments, reclamation bonds, or equivalent mechanisms can promote environmental management from the outset, and a carefully planned waste management

[27] C. Morgan, 'The Privatisation of State Industries—Guidelines on Environmental Liabilities: Third Newsletter of Mining and Environment Research Network', December, 1992.
[28] Loayza (1992).

programme from the start of operations will assist in spreading the costs and reducing potential hazards.

The 'polluter pays principle' only holds if the polluter can survive in order to pay. The capacity to innovate, including capabilities in environmental management, is one key factor in determining a firm's ability to survive, grow, and continue to generate wealth from metals production.

The Role of Technology Transfer in an Environmental Management Policy

Technological and managerial capabilities are not required just to deal with new and emerging technologies, they are also vital to an environmental management strategy using existing technology due to pervasive inefficiencies. Technology transfer and technology partnership through joint venture arrangements is one way to overcome these constraints, particularly in the developing country context, although such strategic alliances are emerging in all the major mineral producing countries. Recent examples of collaborative partnerships in innovation include: Outokumpu and Kennecott, Outokumpu and CODELCO, Cyprus Mines and Mitsubishi, and Comalco, Marubeni Corporation of Japan, and, the Chilean power company, Endesa.

However, there is a need to broaden the common concept of technology transfer to achieve the desired result of a real transfer of environmental management capability. Traditionally technology transfer has meant a transfer of capital goods, engineering services, and equipment designs—the physical items of the investment, complemented by training in the skills and know-how for operating the plant and equipment. Such transfers are often restructured in scope to match the 'step-increments' involved in add-on regulatory–response technical change.

New forms of technology transfer will need to go further to embrace: first, the knowledge, expertise, and experience required both to operate and manage technical change of an incremental and radical nature; and second, the human resource development and organizational changes involved in an overall approach to improve efficiency and environmental management throughout the process route, plant, and facility.

Many trans-national companies play a major role in the mining industry, usually contributing significant managerial and engineering expertise in joint ventures and subsidiaries. They usually limit their contribution in the light of the costs, capabilities, and work involved to fulfil the immediate needs of the specific investment project or physical item of technology transfer. Empirical research on other sectors demonstrates considerable potential to increase those contributions without adversely affecting corporations' strategic control over 'proprietary' technology.[22,29,30] What is required is a strategy of technology and enterprise targeting and a clear set of technology transfer objectives and financial

[29] A. Warhurst, 'Technology Transfer and the Development of China's Offshore Oil Industry', *World Development*, 1991, **19** (8), pp. 1055–1073.

[30] M. Bell, 'Continuing Industrialisation, Climate Change and International Technology Transfer', A report proposed in collaboration with the Resource Policy Group, Oslo, Norway. SPRU Report, December, 1990.

mechanisms to cover the extra costs involved over and above the investment and basic training budget.

There already exists a developed market and a range of commercial channels through which mine operators can purchase capital goods, engineering services, and design specifications; however, the market for knowledge, expertise, experience, and accelerated training programmes is less mature. Bilateral and multilateral agencies and development banks can play a major role in improving this situation. Such an approach was at the heart of the strategy of China's National Offshore Oil Corporation (CNOOC) which targeted specific major oil companies and required them under technology transfer agreements to transfer the capabilities to master selected areas of technology.[29] Another interesting example is the Zimbabwe Technical Management Training Trust. It was founded by RTZ in 1982 with the aim of training South African Development Community (SADC) professionals in technical management and leadership. It effectively combines academic and on-the-job training in both home and overseas operations, providing possibilities for accelerated managerial learning by being exposed to on-the-job problem-solving situations with experienced colleagues in a range of challenging technical scenarios. Although smaller in scale, this is a similar strategy to that designed by CNOOC to ensure that its trainees worked alongside experts, in different trans-national oil companies, in situations which ensured a 'mastery' of the technology rather than the knowledge of how to create it. RTZ absorbs the entire cost of the training and related MSc scholarship programme at the City of London Business School.

It would be quite feasible to build similar in-depth training programmes, concentrating on human resource development in environmental management, into many of the proposed and prospective mineral investment projects throughout the world. Investors and technology suppliers could be selected in part on the basis of their proven environmental management competence and their willingness to transfer it. It cannot be over-emphasized that all technology transfer and training has a set of costs for the supplier and these must be covered to ensure optimal results. The danger of not budgeting for these costs would be to resort back to a training programme in operational skills rather than in technology 'mastery' skills. The corporate partners, the government, donor agencies, or development banks could assist in finance. However, negotiating power regarding the precise objectives and scope of the programmes would be greater if governments, organizations, or firms contributed financially. This paper suggests that it is the capacity to effect technical change, not just the skill to operate an item of environmental control technology, which will ultimately determine the success with which firms build up competence in environmental management.

Moreover, referring back to our case study examples of best practice environmental management, the technical changes introduced illustrate the myriad of intangible practices which constitute sound environmental management. Of relevance to technology transfer policy is the fact that it is not the utilization of one specific technique, but rather a combination of technology, managerial approaches, work-place practices, and regulatory and monitoring frameworks, which explains the parallel achievements of improving economic efficiency and environmental performance. Most of this capacity would not be secret or proprietary to the firm. It is more a question of knowing what to ask for and then being prepared to pay the cost of the resources, time, and effort required for its transfer. This reinforces the need for mining enterprises to build up a range of technological and managerial capabilities as well as workers' skills. This also explains why in the developing country context the simple act of acquiring a new item of technology or an add-on piece of pollution control will not automatically lead to an expected increment of pollution reduction. The transfer of environmental technology does not lead to a transfer of environmental management capabilities unless training and broader human resource development programmes are built into the investment and an overall approach to improve efficiency and good housekeeping at the plant site is adopted. In the developing country context such change is best brought about when stimulated by donor agency requirements supported by credit, government commitment, and the participation in the investment project of a partner with proven competence in environmental management elsewhere.

A further public policy implication suggested by the preceding analysis is the need for a two-tier environmental management policy. There is one set of challenges for the on-going minerals industry, which must encompass the findings above regarding production inefficiency and its environmental consequences and the clean-up requirements on mine shut-down and plant decommissioning. Another set of challenges concerns the policy need to build environmental management into investment and expansion projects from the outset, which requires negotiation at the earliest stage with operators, equipment suppliers, and credit sources.

The public policy challenge is, therefore, how to keep firms sufficiently dynamic to be able to afford to clean up their pollution and generate economic wealth through innovation and sustainable environmental management practices. The achievement of improved production efficiency and environmental management, particularly in developing countries, will in turn be dependent upon the extent to which far-reaching technology transfer and training clauses are built into the joint ventures and new investment arrangements which characterize the industry, and whether banks, donor organizations, and governments demonstrate responsibility by providing the appropriate lines of credit and technical assistance in support of such objectives. Environmental regulation would be one element of that policy and would provide the goal posts for site-specific best practice in environmental management; and, technology policy to promote technical change through technology transfer and human resource development would lie at its heart.

Acknowledgements

The author would like to acknowledge gratefully the kind assistance of Allison Bailey, Gill Partridge, and Hilary Webb in the preparation of this article. Support for the research reported here was provided by the John D. and Catherine T. MacArthur Foundation. Parts of this work build on a more detailed study entitled Environmental Degradation from Mining and Mineral Processing in Developing Countries: Corporate Responses and National Policies, A. Warhurst, 1991, to be published as a book by the OECD Development Centre, 1993. Parts of this paper were presented in the John M. Olin Distinguished Lectureship Series in Mineral Economics at the Colorado School of Mines in December, 1992 which are being published in 'Mining and the Environment: International Perspectives on Public Policy', edited by Roderick G. Eggert, Resources for the Future, Washington, DC, 1994.

Subject Index